Food Safety and Product Liability

Raymond O'Rourke
LLB, Barrister
with Mason, Hayes & Curran, Dublin

Palladian Law Publishing Ltd

© Raymond O'Rourke 2000

Food Standards Act 1999: Crown Copyright 1999 with the permission of the Controller of Her Majesty's Stationery Office

Published by
Palladian Law Publishing Ltd
Beach Road
Bembridge
Isle of Wight PO35 5NQ

www.palladianlaw.com

ISBN 1 902558 22 7

Typeset by Heath Lodge Publishing Services
Printed in Great Britain by The Cromwell Press Ltd

To Olive

· Contents ·

· Preface ·

"Early and provident fear is the mother of safety." Edmund Burke (1729-1797)

Burke's words are prophetic for those living in the United Kingdom in the last number of years when it comes to the subject of food safety. In the past year it has been hard to escape issues of food safety in the UK media, such as Dioxin contamination in Belgium, the safety of genetically modified foods and the quarrel between France and Britain over the lifting of the EU ban on British beef. Food scares such as those in relation to BSE, GM foods and benzene in carbonated water, have created "fear" amongst consumers about the safety of the food they purchase. No other issue has the potential to create hysteria in the media and amongst the general public as does a food scare.

In 1999 the Blair Government which prides itself on being in tune with the views of consumers found itself caught, for many months, between defending the hi-tech biotechnology industry whilst trying to allay consumers' fears about the safety of GM foods. The previous Conservative Government had been rocked to its foundations by the BSE crisis and John Major has subsequently admitted that it was the most intractable issue for his Government, even more difficult than divisions in his Government over Europe. What this demonstrates is that food safety and issues concerning food law are gaining more importance for the Government, the food industry and consumers every year.

Food law is now classed as a particular "specialisation" within the legal profession but unfortunately there is very little literature available which is easily accessible for legal practitioners. I am very happy therefore to be involved with my publishers who have been at the forefront of providing food law texts for the legal profession, as well as for those working in the food industry, environmental health and trading standards officers. This book, in many ways, is a companion volume to my earlier book, *European Food Law*, also published by Palladian Law Publishing. In that book, I discussed all the main changes introduced to the EU regulatory regime for foodstuffs following the BSE crisis. The present book looks in detail at how these changes have impinged on the existing food law in the United Kingdom, as well as referring to the situation in Ireland.

The book begins with a detailed discussion of the Food Safety Act 1990, which was a pioneering piece of legislation when introduced over 10 years ago. The Food Safety Act 1990 is seen as the major framework piece of UK food law that exists in conjunction with all the EU food legislation which must be incorporated into UK law. The chapter also includes a detailed discussion on the case law involving the Food Safety Act 1990. From this survey it can be seen that there is no actual sentencing pattern to the type of fines imposed on food companies for producing unsafe food. There are always differences in sentencing policy between different areas of the country, but the vast differences in the level of fines imposed by the courts does seem to show that magistrates in particular are unsure about the true reasoning behind this legislation. The "due diligence" defence available under the Food Safety Act 1990 has also caused some difficulties in court proceedings. Although the Act has operated reasonably effectively, in the ensuing decade the United Kingdom has witnessed a number of food scares such as salmonella in eggs and the BSE crisis, which undermined the entire UK regulatory regime for foodstuffs. It is not surprising therefore that the present Blair Government has tried to tackle many of the regulatory "loopholes" which have emerged following these food scares.

Questions of food safety naturally create issues of liability for the food industry, retailers, restaurants or local authorities. In relation to liability the book discusses the enactment of the EU Product Liability and General Product Safety Directives into UK and Irish law. The Product Liability Directive has now been extended to cover primary agricultural products. Since consumers are paying ever more attention to the safety of food products, there is certain to be an increase in the number of legal proceedings seeking redress against agricultural producers. The European Commission has also published a Green Paper on Product Liability and a Review Study of the General Product Safety Directive. These initiatives are likely to introduce further amendments to these Directives in the coming years.

Issues of liability in relation to genetically modified foods are also covered in the book. In particular, the Genetically Modified Foods and Product Liability Bill 1999 is discussed. This is an area that is likely to be legislated for in the future. Equally a number of "class action" cases against bio-tech companies have begun in the United States and their outcome will have an influence on whether similar cases will taken in the UK courts.

The book then looks at specific areas of food law such as hygiene, labelling, additives and compositional standards for particular foodstuffs (chocolate, natural mineral waters, and baby foods) where the majority of UK laws consist of statutory instruments enacting EU legislation. All of these laws create obligations and liabilities for the food industry and enforcement authorities. Although prosecutions can be taken against unscrupulous operators under all these Statutory Instruments, the majority of food law prosecutions are taken under various sections of the Food Safety Act 1990, which create specific food law offences.

The book concludes by covering the Food Standards Act 1999, establishing the Food Standards Agency, which comes into operation in April 2000. A complete text of the Food Standards Act 1999 is contained in an Annex to assist the reader in understanding the important new powers which have been given to this agency. The Blair Government on coming to power promised to tackle the whole issue of food safety and in January 1998 published a White Paper on the establishment of a Food Standards Agency. This White Paper was subsequently used as the basis for a parliamentary bill, which finally was enacted into law as the Food Standards Act 1999 in November 1999.

As the Food Standards Act 1999 specifies, the main objective of this new agency is "*to protect public health from risks which may arise in connection with the consumption of food (including risks caused by the way in which it is produced or supplied) and otherwise to protect the interests of consumers in relation to food*". The agency will be independent and have wide-reaching powers in the formulation and drafting of food legislation. Its scope will embrace the entire food chain with a remit to cover everything from food labelling and standards to negotiating about food policy issues on the United Kingdom's behalf at EU and international level. It will provide advice and information to ministers, the food industry and the public at large on food safety and nutrition issues. Finally, it has particular audit powers to monitor and enforce existing UK and EU food law.

While all these developments concerning food law are taking place in the United Kingdom, the European Commission has recently published a White Paper on Food Safety (January 2000). The Commission states in this document that the most appropriate response to the need to guarantee the highest EU food safety standards is the establishment of an independent European Food Authority. The White Paper on Food Safety also includes a comprehensive range of over 80 areas where European food law needs to be amended and improved. The list includes a general food law Directive, which will embody the responsibilities and obligations to be placed upon all operators in the food chain from farmers to food retailers.

The future of UK food law is therefore very much dependent on the outcome of these initiatives and legislative proposals emerging from Brussels. There are likely to a number of "turf wars" between the Food Standards Agency and the European Food Authority over the question of which agency has competence in a particular area of food law. Likewise, the forthcoming EU general food law Directive may compliment or may come into conflict with existing UK law, in particular the Food Safety Act 1990, which is seen as the framework UK legislation covering the obligations on all those dealing with food such as producers, manufacturers, retailers and enforcement agencies.

Food law is certain to become even more important in the coming years therefore lawyers, the food industry and enforcement agencies will need to be well versed in all aspects of UK food law as described in this book. The success of all these new initiatives in the area of food law will be demonstrated when consumers no longer fear for the safety of the food they eat or purchase. The road ahead is certain to be eventful, but if these initiatives fail, not only will consumers continue to have distrust in the safety of food, but they will also have lost confidence in the ability of the UK regulatory authorities to protect the health of its citizens, which is one of the major responsibilities of any government.

Raymond O'Rourke
Dublin
February 2000

· Table of Cases ·

· Table of Statutes ·

[page references in bold denote where text of Food Standards Act 1999 is reproduced in Appendix 1]

· Table of Statutory Instruments ·

Ireland

· Table of EU Legislation ·

Regulations

Chapter 1

. Framework Food . Legislation

Summary

- Food Safety Act 1990 – framework law
- Food safety offences (sections 7 and 8, 1990 Act)
- Consumer protection offences (sections 14 and 15, 1990 Act)
- Improvement and prohibition notices
- Emergency prohibition and emergency control orders
- Defences available under 1990 Act – due diligence
- Prosecutions and penalties under 1990 Act
- Case law
- Food safety law in Ireland

1.1 **Introduction**

The enactment of laws concerning issues of food safety and food quality has a long and chequered history in the United Kingdom. The first English food law provisions can be traced back to those established by trade guilds in the form of voluntary self-regulation aimed at combating the adulteration of food. Such laws ensured that consumers would be provided with the correct weight of bread and would avoid the possibility of obtaining unsound beer. Later in the 18th and 19th centuries various comprehensive acts of parliament were introduced once again to prevent the adulteration of specific products such as tea, coffee and bread. Towards the end of the 19th century laws were introduced obliging local authorities to appoint public analysts and inspectors whose powers were to procure samples for analysis and to obtain sufficient evidence to convict guilty parties for the adulteration of food. In the 20th century various Food and Drugs Acts were enacted by the UK Parliament aimed at safeguarding health and preventing fraud and deception. Nowadays although the main controlling provisions of

UK food law are contained in the Food Safety Act 1990[1], any food regulations made prior to this Act under previous Food Acts are now regarded as enforceable under the Food Safety Act regime. Naturally, as the title of the Act implies, the Food Safety Act 1990, contains a number of detailed provisions whose aim is to increase the safety of food and food products throughout the entire food chain.

1.2 **Food Safety Act 1990**

Section 1 of the Food Safety Act broadly defines food, giving a non-exhaustive list. Food includes:

- drink;
- articles and substances of no nutritional value, which are used for human consumption;
- chewing gum and other products of a similar nature; and
- articles and substances used as ingredients in the preparation of food.

Those "food" products not included within the ambit of the Act include:

- live animals or birds, or live fish which are not used for human consumption while they are alive;
- fodder or feeding stuffs for animals, birds or fish;
- controlled goods within the meaning of the Misuse of Drugs Act 1971; and
- other such medicinal products within the meaning of the Medicines Act 1968.

Section 1(3) of the Act defines a number of terms including "business", "commercial operation", "food business", "food premises", "food source" and "premises". A **business** includes the undertakings such as canteens, clubs, schools, hospitals or other such institutions whether or not carried on for profit reasons. **Commercial operations** in the food area include any of the following:

- selling, possessing for sale and offering, exposing or advertising food for sale;
- consigning, delivering or serving food for sale;

---

1   Chapter 16.

- preparing for sale or presenting, labelling or wrapping of food for the purpose of sale;
- storing or transporting food for the purpose of sale;
- importing and exporting any food product.

"**Food business**" means any business in the course of which commercial operations with respect to food or food sources are carried out. "**Food premises**" means any premises used for the purposes of a food business. A "**food source**" means any growing crop or live animal, bird or fish from which food is intended to be derived (whether by harvesting, slaughtering, milking, collecting eggs or otherwise). "**Premises**" includes any place, any vehicle, stall or moveable structure or for such purposes may be specified in any order made by UK ministers, any ship or aircraft of a description so specified.

The Act will also apply to any food which for the purpose of advertisement or in the furtherance of any trade or business is offered as a prize or reward or given away and in the same way it is understood that such food is seen to be equivalent to food for sale.

Under section 3(2) of the Food Safety Act it is presumed that food is intended for human consumption unless the contrary is proven. It is for the defendant to prove to the court that food or articles or substances capable of being used in the consumption or preparation of food, kept on the food premises were not intended for sale or manufacture for human consumption.

Ministerial responsibility and powers to issue regulations

Section 4 of the Food Safety Act indicates which ministers have tasks under the Act. The Minister of Agriculture, Fisheries and Food and the Secretary of State for Health all have responsibilities. In the original Act the Secretary of State for Scotland and Wales were also specified as having responsibilities in the enforcement of the Food Safety Act, but with new devolved powers to the Scottish and Welsh Parliaments this will change and the Ministers for Health and Agriculture both in the Welsh and Scottish executives will now have powers to enforce elements of the Food Safety Act in their own jurisdictions. The ministers specified under the Food Safety Act have the power to issue regulations on several subjects of which the most important are mentioned below.

- Section 16 of the Act gives ministers powers to make regulations on various matters to ensure adequate standards of food safety and consumer protection such as food composition, food safety, food labelling and hygiene.

- Section 17 permits ministers to make regulations in order to fulfil any European Union obligations concerning food and to secure that directly applicable EU provisions relating to food are administered, executed and enforced.
- Under section 19 ministers can make regulations for the registration and licensing of food premises.
- On the basis of section 45 of the Act ministers may also issue regulations as regard particular charges to be imposed by enforcement authorities in the food area and may also under section 48(2) make regulations or orders exercisable by statutory instrument. Before making any regulation, (except under ss 17 and 18 of the Food Safety Act) the ministers have to consult with organisations which appear to them to be representative of interests likely to be substantially affected by the proposed legislation.

Offences under Food Safety Act 1990

The Food Safety Act 1990 distinguishes between two kinds of offences, food safety offences (ss 7 and 8) and consumer protection offences (ss 14 and 15). In addition, the Food Safety Act contains the general offence of obstructing an officer (s 33). Offences can be committed by natural persons and bodies corporate. Under section 36 of the Food Safety Act the manager, director, secretary or similar official of a body corporate can be liable for prosecution where it is proved that he/she negligently consented to or connived at the alleged offence. Not only will the body corporate be guilty of the offence but so will these individuals.

Food safety offences

The two specific food safety offences laid down in the Food Safety Act relate to rendering food injurious to health and selling food not complying with the food safety requirements. Section 7 of the Act makes it an offence for any person who renders any food injurious to health to be prosecuted if the food has undergone one of the following operations, namely:

"(a) Adding any article or substance to the food;
 (b) Using any article or substance as an ingredient in the preparation of the food;
 (c) Abstracting any constituent from the food; and
 (d) Subjecting the food to any other process or treatment, with intent that it shall be sold for human consumption.".

The section specifies further that in determining whether any food is injurious to health, regard will be had not only to the probable effect that food may have on the health of a person consuming it but also to the probable cumulative effect of food of substantially the same composition on the health of a person consuming it in ordinary quantities.

The second food safety offence laid down in the Act refers to selling food that fails to comply with food safety requirements. For the purposes of a section 8 offence food fails to comply with food safety requirements if:

"(a) It has been rendered injurious to health by means of any of the operations referred to in section 7, (such as the addition of any article or substance to the food etc);

(b) It is unfit for human consumption; or

(c) It is so contaminated (whether by extraneous matter or otherwise) that it would not be reasonable to expect it to be used for human consumption in that state."

It is important to mention that section 8(3) of the Food Safety Act points out that any food which fails to comply with food safety requirements, whether in terms of a batch, a lot or consignment of food of the same class or description, would be generally presumed to be covered by a section 8 offence. Until the contrary is proved all of the food of that batch, lot or consignment would fail to comply with these food safety requirements. The decision as to whether to outlaw a complete batch, lot or consignment of food would be taken by an authorised officer of the food authority following an inspection under section 9 of the Food Safety Act. It is also important to note that the Food Safety Act does not define the notion of contamination, rather in general it is interpreted widely covering any foreign body, mould, pesticides, residues, heavy metal presence, mite and similar infestation, radioactive contamination and the unauthorised use of additives affeting the safety of food. Contamination can be of a physical, chemical or biological nature. Whether an offence under section 8 has been committed depends on the circumstances and practice for which no exact rules can be determined. In addition, whether food is unfit for human consumption or not is largely a question of fact and it is for the courts to decide in the final analysis.

Consumer protection offences

In relation to consumer protection offences, under section 14 of the Act it is an offence to sell any food which is not of the nature or substance or quality demanded by the purchaser. "Sale" means sale or other supply for human consumption. "Not of the nature" refers, for instance, to selling a variety of food not requested, *e.g.* selling a stew containing too little meat or with the variety of meat not labelled. Selling food not of the substance demanded may occur where food is being substituted by the addition of an adulterant, such as chicory in coffee. A foodstuff is not of the quality demanded where it falls short of what is reasonably expected. Under a section 14 offence the burden of proof rests with the prosecution for cases involving contamination of food, for example the presence of foreign bodies or micro-organisms. There is a clear overlap between a section 8 and a section 14 food safety offence.

Paragraph 18 of the Code of Practice No 1 in relation to the Food Safety Act recommends that cases involving contamination involving the micro biological quality of food should be prosecuted under sections 7 and 8 wherever possible.

Under section 15 of the Act it is an offence for a person who gives or displays for sale any food with a label which:

> "(a) Falsely describes the food; or
> (b) Is likely to mislead as to the nature, substance or quality of the food."

Further, under section 15(2) any person who publishes or is a party to the publication of an advertisement which falsely describes any food or is likely to mislead as to the nature of the food or substance or quality of any food likewise will be guilty of an offence.

Under section 15(3) any false presentation of a food product likely to mislead the consumer likewise is an offence. The difference between false and misleading is that a label is false if there is a clear factual misstatement; the label is misleading if the label is false only by inferring or omitting something of importance.

Enforcement powers

Under section 9 of the Act an authorised officer of a food authority may at all reasonable times inspect any food intended for human consumption which has been sold or is offered for sale in order to ascertain whether that food product fails to comply with specific food safety requirements or is likely to cause food poisoning.

Under section 9(3) the authorised officer may either:

"(a) Give notice to the person in charge of the food that, until the notice is withdrawn, the food or any specified portion of it –

(i) is not to be used for human consumption; and
(ii) either is not to be removed or is not to be removed except to some place specified in the notice; or

(b) Seize the food and remove it in order to have it dealt with by a Justice of the Peace."

Where an authorised officer exercises the powers conferred by subsection (3)(a) above he shall as soon as is reasonably practicable, or at least within 21 days, determine whether or not he is satisfied that the food complies with food safety requirements and if he is so satisfied will withdraw the notice and if he is not so satisfied shall seize the food and remove it in order to have it dealt with by a Justice of the Peace. Obviously, if the authorised officer decides to seize the food and have the issue dealt with by a Justice of the Peace he must inform the person in charge of the food of his intention to have it dealt with in this way.

In connection with any particular court hearing under section 9(6) of the Act if the court decides that the food does indeed fail food safety requirements it must decide how it should be destroyed or disposed of and any expenses incurred in connection with such destruction or disposal must be defrayed by the owner of the food. On the other hand, if the court decides that the food product is indeed safe then it must ensure that the food authority shall compensate the owner of the food for any depreciation in its value resulting from the action taken by the authorised officer. Any disputed question as to the right to or amount of any compensation payable in this latter context under the Act shall be determined by arbitration.

Improvement notices

The authorised officer of a food authority also has powers under section 10 of the Food Safety Act if he has reasonable grounds for believing that the proprietor of a food business is failing to comply with any food safety regulations he may by a notice served on the proprietor introduce what is referred to as an "improvement notice". This improvement notice will:

"(a) State the officer's grounds for believing that the proprietor is failing to comply with the regulations;

(b) Specify the matters which constitute the proprietor's failure so to comply;

(c) Specify the measures which in the officer's opinion the proprietor must take in order to secure compliance; and

(d) Require the proprietor to take those measures or measures which are at least equivalent to them, within such period (not being less than 14 days) as may be specified in the notice."

Any person who fails to comply with such an improvement notice shall be guilty of an offence under the Act.

Prohibition notices

Under section 11 of the Food Safety Act there is provision for what is termed a prohibition order which will be imposed by the courts at the time of conviction in relation to an offence, *e.g.* section 7 or 8 offence under the Food Safety Act. Such an order will be imposed if two specific conditions are fulfilled.

(1) A business owner must have been convicted under any of the regulations to which section 11 applies, *i.e.* the non-observance of hygienic conditions and practices in connection with the carrying out of commercial operations with respect to food or food sources.

(2) The court must be satisfied that the health risk condition has been fulfilled with respect to that business, *i.e.* that if there is a risk of injury to overall public health arising from the use of any process or treatment, premises or equipment.

The type of examples would be serious infestation by rats, mice, cockroaches or other vermin, serious damage effects of flooding to premises, use of defective equipment, serious risk of cross-contamination or inadequate temperature controls. Interestingly, under section 11(4)(b) of the Act the court has a discretionary power to impose a prohibition order on the proprietor participating in the management of any food business or any food businesses of a class or description specified in the order for the foreseeable future. The intent of this section of the Act is to prevent a convicted proprietor or manager of a food premises from re-opening almost immediately in a new food premises. The prohibition on managing a food business under section 11(4) is rigorously applied in practice and LACOTS (Local Authorities Co-ordinating Body on Food and Trading Standards) has published a list of some 15 persons who have been convicted under this prohibition in 1997, 1998 and 1999.

As soon as practicable after the institution of the court order, the enforcement authority must serve a copy of the order on the proprietor of the business. If the order specifically refers to the premises, it will be fixed prominently on those premises and must not be removed or obscured. The order ceases to have effect when the enforcement authority is satisfied that the "health risk" condition no longer applies following the various steps taken by the owner. A prohibition order preventing the proprietor or manager from participating in any food business as specified under section 11(4) can only be lifted upon the direction of the court. In order to avoid formal prohibition measures, the owner of a food business may offer to close voluntarily. If the authorised officer accepts such an offer under the Codes of Practice he must obtain written confirmation of the proprietor's offer to close and not to open without his specific permission. The authorised officer is asked to ensure that frequent checks are made of the premises to ascertain that they are not re-opened.

Emergency prohibition orders

Another issue in relation to enforcement powers concerns emergency prohibition notices and orders which can be issued under section 12 of the Food Safety Act. This section provides authorised enforcement officers with powers to issue emergency prohibition notices and orders to deal with circumstances which pose an imminent risk of injury to health. This power was introduced in section 12 to cover potentially high risk situations, for instance, if the condition of the food premises appeared to carry a high risk of causing an outbreak of food poisoning within the forthcoming days or a defective food manufacturing process could result in an incidence of botulism or food poisoning. Once the authorised officer is satisfied that there is such a health risk he must obtain clearance from the magistrates' court to impose such an order.

Under section 12(3) such an officer shall not apply for an emergency prohibition order to the magistrates' court unless at least one day before the date of such application he has served notice on the proprietor of the business of his intention to apply for the order. The authorised officer must ascertain the risk of injury to health, whether in terms of the purposes of the food business or processes used in that food business or the construction of any such food premises in line with section 11(2) and (3) of the Act. Both an emergency prohibition notice and order cease to have effect upon the issue by the enforcement authority of a certificate that it is satisfied that the proprietor has taken sufficient measures to secure that the business no longer poses a health risk.

There is no right of appeal against an emergency prohibition notice, but the notice may be reviewable under the compensation provisions contained in section 12(10) of the Act. There is though a right of appeal to a magistrates' court by way of complaint against refusals of an enforcement authority to issue a certificate of satisfaction that the proprietor has taken sufficient measures to ensure that his business no longer poses a health risk (s 37(1)(b) Food Safety Act 1990). Where the magistrates' court refuses to allow the appeal, the appellant has the right to appeal to the Crown Court (s 38(a) Food Safety Act 1990). There is also a right of appeal to the Crown Court against a decision by the magistrates to issue an emergency prohibition order (s 38(b) Food Safety Act 1990). In certain situations, the owner of a food business may offer to cease trading on a voluntary basis. Most food authorities are prepared to accept such an offer.

Emergency control orders

Under section 13 of the Food Safety Act 1990 the Minister is given powers to issue emergency control orders, for example to halt commercial operations in respect of food which may involve an imminent risk of injury to health. The purpose of this power is to allow the minister to deal with emergency situations, which obviously are too extensive or too serious to be dealt with by local food authorities, for example in the case of contamination of packaged food which has been distributed across the country.

Under section 13(5) the Minister may give any such directions as appear to him to be necessary or expedient for the purpose of preventing the carrying out of commercial operations with respect to any food, food sources or contact materials or anything which appears to him to be necessary or expedient for that purpose. Interestingly, as of December 1999 the power of the Minister to issue an emergency control order has not been exercised.

Defences

The Food Safety Act 1990 follows a strict liability regime, whereby the burden of proof on the prosecutor in such cases is to prove any such offence to the criminal standard, that being "beyond reasonable doubt". Basic elements required for criminal liability in English criminal law are a mental element *mens rea* and a physical element *actus reus*. For strict liability offences there is an absence of recklessness or intention (*mens rea*). Strict liability, therefore, is usually found in regulatory legislation, for instance food law offences may be punished by a fine and/or imprisonment. On the other hand, the standard of proof for the defendant is the lesser civil standard of the "balance of probabilities". When a defendant wants to depend on a defence, such as those contained in the Food Safety Act, he must prove it by applying the civil standard. The burden of proof on the prosecutor remains the criminal standard "beyond reasonable doubt", but when the prosecutor fails to prove his case there is no need for the defendant to prove a defence. In order to prevent unfair situations occurring, the Food Safety Act 1990 contains three main defences, in relation to any offences brought against a food manufacturer/retailer under the Act.

Offences due to fault of third person

The first defence involves offences due to the fault of another person. Under section 20 of the Act where it can be proved that an offence is due to the act or default of some other person, that other person shall be guilty of the offence and an enforcement authority can bring proceedings against either the principal or the other person or both. The aim of this defence is to ensure that the person actually responsible for committing the offence is prosecuted. This defence could be relied upon, for example, in a case where a "foreign body" was found in tinned food. Technically the retailer who sold the foodstuff to the consumer would have committed an offence under section 14 of the Food Safety Act. However, where it could be proved after investigation that the foreign body came into the tinned foodstuff as a result of some fault in the production process, section 20 would thereby enable the prosecution to charge the manufacturer of the tinned food product instead of the retailer and section 20 would obviously give the retailer a defence in any charge brought against him.

Due diligence

A second and important defence which can be relied upon under the Act is the defence of due diligence (s 21 Food Safety Act). This defence was specifically introduced into food law for the first time by the Food Safety Act 1990. The Act states that it is a defence for a person to prove that:

> "He took all reasonable precautions and exercised all due diligence to avoid the commission of the offence by himself or by a person under his control." (s 21(1))

The aim of this defence therefore is obviously to prevent the conviction of a defendant who did all he reasonably could do to prevent a breach of the law. Whether this defence has been satisfied is a question of fact, to be determined in every single case. The onus of proof lies with the defendant on the basis of the civil standard of proof, *i.e.* on the "balance of probabilities".

The expressions "all reasonable precautions" and "all due diligence" are not specifically defined; however, the general understanding (based on case law) is that the expression "**all reasonable precautions**" means that a control system must be established and "**due diligence**" means that steps are taken to ensure that the system is working satisfactorily. Such a system includes compliance of materials used with the law, checked at every stage of the food production process, cleanliness and hygiene of buildings, machinery and instruments, training of staff, dealing with complaints, staff management etc. It is closely associated with the quality assurance systems provided by ISO 9000/HACCP and companies that have such systems in place are likely to be covered in terms of "due diligence" for legal purposes. Part of the defence of "due diligence" may be to show that someone else was at fault. Under section 21(5) of the Act in such a case, the person charged should provide this information to the court at least seven days before the hearing or if the defendant has already appeared before the court, within one month of that appearance. However, anyone intending to make such a claim must inform the prosecution in advance.

Publication in course of business

A third defence involves a defence of publication in the course of business (s 22 Food Safety Act). In this situation regarding the advertisement for sale of any food product, it should be a defence for the person charged of an offence under the Food Safety Act to prove that he is a person whose normal business is to publish or arrange for the publication of advertisements and that he received the advertisement for sale of the food product in the ordinary course of his business and did not know and had no reason to suspect that its publication would amount to an offence under the Food Safety Act 1990. The burden of proof is on the advertiser to prove the statutory defence on the civil standard, *i.e.* "balance of probabilities".

Prosecutions

It should be noted that under section 6(5) of the Food Safety Act 1990, enforcement authorities may institute proceedings under any provisions of the Food Safety Act 1990 or any statutory legislation enacted under it. The decision of the enforcement authorities to prosecute may be the result of a situation discovered following an inspection visit by an authorised officer or merely a consumer complaint. However, it is specified that legal proceedings under the Food Safety Act 1990 should be brought without unnecessary delay. Although a food authority has a duty to enforce the provisions of the Food Safety Act 1990, there is no automatic obligation upon them to continually prosecute for food safety offences and any time they decide to prosecute they must take the general public interest into account before deciding on this outcome. More specific guidance on this matter for enforcement authorities is laid down in a number of codes of practices which have been introduced under section 40 of the Food Safety Act 1990 which provides for such codes to be enacted. These codes, for example, mention that the enforcement authority (prosecutor) should consider the seriousness of the offence, the previous history of the party concerned, the willingness of the party to prevent any recurrence of the problem and any explanation offered by the affected food company. Specific time limits are specified under section 35 of the Food Safety Act 1990 for prosecutions. This section specifies that no prosecution for an offence under the Food Safety Act 1990 is punishable after the expiry of:

(a) three years from the commission of the offence, or
(b) one year from its discovery by the prosecutor (enforcement authority),

whichever is the earlier.

Penalties

Section 35 of the Food Safety Act 1990 specifies sanctions for any violations under the Act. For the offence of obstructing an authorised officer of an enforcement authority entering premises to take samples, or inspecting the building to ascertain whether there have been food safety violations (s 33(1)), a person guilty of this offence shall be liable on summary conviction to a fine not exceeding £5,000 or to imprisonment for a term not exceeding three months or to both.

A person guilty of any other offence under this Act shall be liable:

(a) on conviction on indictment, to a fine or to imprisonment for a term not exceeding two years or to both;

(b) on summary conviction, to a fine not exceeding the relevant amount or to imprisonment for a term not exceeding six months or to both.

In relation to the "relevant amount" in the case of an offence under section 7, 8 or 14 of the Food Safety Act 1990 a fine will be to the amount of £20,000 and/or imprisonment up to six months. For all other offences under other sections of the Act the relevant amount will be a fine of £5,000 and/or imprisonment up to six months. Finally, in the case of a conviction on indictment the relevant amount can be an unlimited fine and/or a period of imprisonment of up to two years or to both.

In addition to the fines and terms of imprisonment which may be imposed for food law offences as set out in section 35 of the Food Safety Act 1990, offenders of the Act can be punished in other ways. For example, businesses can be ordered to pay compensation to consumers who have been injured by their food. There are three ways in which this could come about:

(1) If the consumer successfully sues for civil damages under common law.

(2) If the consumer successfully obtains damages under the Consumer Protection Act 1987 (EU Product Liability Directive regime).

(3) If the criminal courts make an order requiring a convicted offender to pay compensation to a person who has suffered loss or injury as a result of a criminal act.

Whilst the fines for individual offences under the Food Safety Act 1990 and the regulations might not seem prohibitive, cumulative fines for numerous offences can be very costly for businesses. Furthermore, the adverse publicity following conviction can be damaging to the image of a food company.

Miscellaneous provisions introduced

A number of statutory instruments have been introduced following the adoption of the Food Safety Act 1990. These include:

(1) Detention of Food (Prescribed Forms) Regulations 1990 (SI 1990 No 2164). These Regulations prescribe forms of notice which may be used in connection with the detention of food under section 9 of the Food Safety Act 1990.

(2) Food Safety (Improvement and Prohibition – Prescribed Forms) Regulations 1991 (SI 1991 No 100). These Regulations prescribe the forms of notice which may be used in connection with improvements notices under section 10, prohibition orders (other than those relating to proprietors) under section 11 and emergency prohibition notices or orders under section 12 of the Food Safety Act 1990.

(3) Deregulation (Improvement of Enforcement Procedures) (Food Safety Act 1990) Order 1996 (SI 1996 No 1683). This order which extends to Great Britain, exercises powers in the Deregulation and Controlling Out Act 1994, to improve enforcement procedures under the Food Safety Act 1990. Before an authorised officer of an enforcement authority serves an improvement notice on the proprietor of a food business under section 10 of the 1990 Act:

   (a) the officer must give the proprietor a written notice stating that the officer is considering serving an improvement notice, the reasons for this and that the proprietor has the right to make representations; and
   (b) the proprietor may within the period specified in the notice for making representations, either make written representations to the officer or if the proprietor so requests within the period specified within the notice for making such request, make oral representations to the officer in the presence of a senior officer of the enforcement authority.

Under this order the officer is obliged to consider any representations which are duly made.

Case law

Section 8 cases

Section 8 of the Food Safety Act 1990 makes it an offence to sell food which is unfit for human consumption. This is a question of fact and in the past a loaf of bread found to contain a used dirty bandage was held to be unfit in *Chibnall's Bakeries* v *Cope Brown*[2] as was a pork pie which contained a harmless black mould under the crust of the pie: *Greig* v *Goldfinch*.[3] Contamination is not defined in the Food Safety Act 1990, but it would include foreign bodies, mould, insect infestation etc. Two 1995 cases consider what evidence is required before there can be a successful conviction for an offence under section 8.

Kwik Safe Group plc v *Blaenau Gwent Borough Council (1995)*[4] concerned the purchase of two packets of crumpets from a local Kwik Save store. The purchaser bought the crumpets in the morning and by the afternoon they were found to be full of mould and in actual fact the goods bought were one day past their best before date. The purchaser brought the crumpets to her local environmental health officer who immediately did tests on them. The shop was fined £250 in the magistrates' court and £250 costs. The shop then appealed on the grounds that there was insufficient evidence before the magistrates to justify conviction as no expert evidence was taken in the magistrates' court. The Crown Court decided that the magistrates were correct in their judgment, as it was beyond reasonable doubt that the crumpets which were bought at the Kwik store in the morning were mouldy at the time of purchase and therefore were unfit for human consumption. The second case, *Ruxtan* v *Haston*[5] was heard before a sheriff court in Scotland. A dairy was charged under section 8 of the Food Safety Act 1990 with having sold to over 50 named persons milk contaminated with E coli 0157 which was unfit for human consumption. The prosecution believed that under the 1990 Act all they had to prove was that the milk was contaminated, but the defence alleged that they were required to prove how and when the contamination took place. The dairy was found guilty and fined £1,000. The court concluded that as the offence was one of strict liability all that the prosecution was required to do was to prove beyond reasonable doubt that the contamination occurred in the milk during the relevant dates with the resulting potential harm to human health.

Finally, *Kellogg's* v *London Borough of Bromley*[6] concerned a customer who found a sharp piece of metal in a packet of Kellogg's Frosties. The company was fined £3,000 with costs of £1,600 under section 8 of the Food Safety Act 1990. Following inspections of

2   [1956] Crim LR 263.
3   (1961) 105 Sol Jo 307.
4   *Justice of the Peace and Local Government Law* 1997 (161) 18, p 431.
5   *Ibid*, 13.
6   (1996) Dukes, *Food Legislation of the UK* (1997) Butterworth, 13.

Kellogg's manufacturing facility it was found that the metal came from a production line machine which broke down. Reference should also be made to the Scots law case of *Errington* v *Wilson*[7] which concerned a condemnation order placed upon a cheesemaker following the finding of listeria in Lanark Blue Cheese which rendered the food unfit for human consumption under section 8 of the Food Safety Act 1990.

Section 10 cases

In relation to improvement notices served under section 10 of the Food Safety Act 1990, the courts have ruled in *Bexley London Borough Council* v *Gardiner Merchant*[8] that the notice must specify the manner of the breach of food safety requirements. In this case it was a breach of the Food Hygiene (General) Regulations 1970 (SI 1970 No 1172). Gardiner Merchant did not meet the requirements of these Regulations, as they did not have conveniently accessible wash-hand basins with hot and cold water in their food premises. As the improvement notice did not specify this fact, Gardiner Merchant successfully appealed a magistrate's decision to the divisional court, even though the magistrate's decision was upheld. In another case involving improvement notices, *Salford City Council* v *Abbeyfield (Worsley) Society*[9], the courts ruled that magistrates had a power to amend an improvement notice. However, in this particular case the amendments could not stand. The case concerned a sheltered home for the elderly, where residents helped in the daily kitchen chores and under food hygiene regulations they should have been provided with adequate kitchen clothing. The magistrates sought to restrict the notice to all "paid" helpers and the divisional court ruled that according to the hygiene regulations they were entitled to do so.

Section 14 cases

Walkers Snack Foods Ltd v *Coventry City Council*[10] raised a number of matters in relation to the investigation by environmental health officers of offences under the Food Safety Act 1990. The case arose over the purchase of a packet of Walkers Crisps, which contained a piece of white plastic. There were three questions dealt with by the court:

---

7   (1996) 2 *Juridical Review*, 153.
8   (1994) 4(2) *Local Government & Law*, 12.
9   *Ibid*.
10 (1998) *Environmental Law Monthly*, August issue.

(1) Environmental health officers were not given all the assistance necessary when they went to inspect Walker's manufacturing facility.

(2) The same officers were refused entry to part of the said premises (an offence under section 33(1)(b)).

(3) Walkers sold food contrary to section 14(1) not of the substance demanded by the purchaser.

Walkers' refusal to allow the environmental health officers access to records and part of the premises because they were from an "outlying" jurisdiction, *i.e.* Coventry rather than Leicester where the factory was located, caused the court to impose fines totalling £10,000 and costs of £28,000. With the benefit of hindsight the company would have fared better if it had taken formal legal advice, rather than acting on its perceived knowledge of the law and in particular the Food Safety Act 1990. In that case, they might have been able to utilise the defence of due diligence, but this failed completely in the court, due to the earlier offences of obstructing authorised officers in their duties.

Other cases involving section 14 offences include *Hatfield Peveral* v *Colchester Borough Council*[11], where a local dairy was charged with selling milk not of the substance required by the purchaser as it contained an earwig. The bottling plant was found to have a particular pest problem after investigation by local environmental health officers. The company was fined £700 with costs of £300. *Rio Trading Co* v *East Sussex Trading Standards Office*[12] also involved a section 14 offence. The firm sold a snack called a "Buzz Bar" to health food shops, which the packaging stated was covered in chocolate. The bar in actual fact was not covered in chocolate but rather a chocolate flavoured coating. The firm was prosecuted under section 14 for selling a snack bar not of the nature demanded by the purchaser as it was not covered by chocolate as stated on the label. The firm was fined £2,500.

Section 16 cases

The issue of emergency prohibition orders (s 16) has been covered in *East Kilbride District Council* v *King*[13] where the owners of a dairy

---

11 Dukes, *op cit*, fn 6.
12 *Ibid.*
13 (1997) 4 *Juridical Review*, 254.

farm were served with an emergency prohibition notice, as there was an "imminent" risk to human health. The environmental health officers within the specified period applied to court to have the notice turned into an order. In that case, the owners of the dairy farm would not be entitled to any compensation, therefore they appealed to the sheriff principal against the emergency prohibition order that was subsequently issued. The sheriff principal judged that no such appeal was provided for in sections 37-39 of the Food Safety Act 1990 under Scots law, but it would be possible under section 38 in England. The anomaly was due to magistrates' courts having "lay" judges rather than the professional judges of Scotland's sheriff courts. The farm owners then appealed to the Scottish Court of Session, but again failed and it was stated such an appeal would also not be possible in England. The judge concluded that this anomaly/confusion should be dealt with by parliamentary legislation

Although a food authority has a duty to enforce the provisions of the Food Safety Act 1990, there is no automatic duty to prosecute for food safety offences and the prosecutor should consider the general public interest before deciding whether or not to prosecute. This was mentioned in an earlier House of Lords case, *Smedleys Ltd* v *Breed*[14] where a food company was prosecuted for supplying a tin of peas that contained a caterpillar which was almost indistinguishable from the peas. The food authority must also prosecute within the time limits provided for in section 34 of the 1990 Act. Under this section the prosecutor must make prosecutions within three years from the commission of the offence or one year from its discovery. This issue was dealt with in *R* v *Thames Metropolitian Stipendary Magistrate ex parte London Borough of Hackney*[15] which concerned a number of recurring offences in a pub/night-club contrary to the Food Hygiene (General) Regulations 1970 (SI 1970 No 1172). Because the proceedings were finally taken against the pub/night-club three and a half years after the date of the alleged offences even though there were recurring food hygiene offences during this period of time, the proceedings were therefore time-barred. The case was appealed to the divisional court who agreed that the case must be closed but there would be no bar to the food authority instituting fresh proceedings which would include much of the evidence put forward in this case.

---

14   (1974) 137 JP 776.
15   *Justice of the Peace & Local Government Law*, 19 Feb 1994, p 117.

Due diligence cases

Finally, the use of the due diligence defence has been covered in a number of court cases. In *Cow & Gate Nutricia Ltd* v *Westminister City Council*[16] the baby food company was prosecuted under section 14 of the Food Safety Act 1990 after a small piece of bone was found in a jar of baby food. Cow & Gate relied on the defence of due diligence in the court hearing. The magistrates ruled that the defence could not relied upon, as the piece of bone was present in the jar of baby food, even so Cow & Gate appealed to the divisional court. The court ruled as the facts of the case were that the piece of bone was present in the jar of baby food this did not deprive Cow & Gate of the defence of due diligence as the magistrates' court implied. The court ruled that the statutory defence of due diligence must be considered in the context of the facts of the case and not rejected as inapplicable because of the facts. In *Carrick DC* v *Taunton Vale Meat Traders*[17] the defendant had consigned to another person meat for sale which was unfit for human consumption. The beef in question was examined by a meat inspector at the defendant's premises and was pronounced "fit to be consigned", but on arrival at the premises of the co-signee it was found to be unfit for human consumption. There was no question that the carcass could have deteriorated during transit, therefore the error was the negligence of the inspector. The defendants who ran a slaughterhouse sought to use the due diligence defence, but the prosecution argued that they should have their own system of checks on carcasses and should not rely entirely on the inspector's reports. The court ruled that the defendants had done all in their power to ensure the food was safe and they could rely on the due diligence defence.

Registration of food premises

Under section 19 of the Food Safety Act 1990, ministers may issue regulations to require the registration of food premises. The Food Premises (Registration) Regulations 1991 (SI 1991 No 2825) were introduced to require the registration of nearly all food premises in the United Kingdom, encompassing premises used for the purpose of a food business on five or more days in any period of five weeks except, for instance, for slaughter houses and meat product plants

---

16　*The Independent*, 24 April 1994.
17　(1994) 58(4) *Journal of Criminal Law*, 325.

which have been approved under other regulations. It is the proprietor of any food business who has the obligation to register and also to notify any changes in the nature of the food business. The proprietor shall apply in writing using the model form printed in the 1991 Regulations for registration with the registration authority in whose area the premises are situated which in practice is the local district council in most cases.

The 1991 Regulations specify that the registration authority shall register the premises within 28 days. By means of registration, food authorities are informed about how many and which food businesses operate within their area so that they can inspect new businesses early on and thereby target inspection, sampling and enforcement resources more effectively. No fee is payable for registration and the registration authority has no powers to refuse, suspend, revoke or attach any conditions to registration. Under the regulations it is an offence not to register and any person who contravenes these provisions shall be guilty of an offence triable summarily and liable on conviction to a fine not exceeding £5,000. There is no element of control in this registration system as every business registers as of right, but it is an offence not to register. Any person who furnishes information which he knows to be false or intentionally or recklessly discloses to another person particulars supplied to a registration authority under these regulations will be guilty of an offence triable summarily and liable on conviction to a fine not exceeding £5,000. It is a defence for a person charged with regard to disclosure of particulars to show that he did not know and had no reasonable grounds to suspect that the person to whom he disclosed the information was not a person to whom such disclosure might be lawfully made.

1.3 **Basis of food law in Ireland**

The foundation of food law in Ireland was the passing of the Sale of Food and Drugs Act 1875, which is still on the statute books and technically remains in force. The Act lays down the fundamental basis of all food law in a number of sections. For example, section 3 makes it an offence to render food injurious to health; section 6 requires that "no person shall sell to the prejudice of the purchaser any article of food which is not of the nature, substance and quality demanded" and section 8 provides that in cases other than health, no offence is committed if due notice be given to the purchaser.

Health Act 1947

In more recent times the Health Act 1947, laid down various provisions relating to offences in the food area. Section 53 of the Act defined food as including every article used for food or drink by man other than drugs or water and:

> "(a) any article which ordinarily enters into or is used in the composition or preparation of human food,
>
> (b) flavouring matters, preservatives and condiments,
>
> (c) colouring matters intended for use in food, and
>
> (d) compounds or mixtures of two or more foods."

Section 54 of the Health Act 1947, has subsequently been amended by the European Communities (Health Act 1947 Amendment of Sections 54 and 61) Regulations (SI 1991 No 333). Under the new provisions the Minister for Industry and Commerce and the Minister for Agriculture and Food may make regulations for:

> "(a) the prevention of danger to the public health arising from the manufacture, preparation, importation, storage, distribution or exposure for sale of food intended for sale for human consumption,
>
> (b) the prevention of contamination of food intended for sale for human consumption,
>
> (c) the prohibition and prevention of the sale or offering or keeping for sale of:
>
> (i)   articles of food intended for human consumption;
> (ii) living animals intended for such food;
> (iii) materials or articles used or intended for use in the preparation or manufacture of such food which are diseased, contaminated or otherwise unfit for human consumption,
>
> (d) the protection of consumer interests."

A person who contravenes section 54, for example, provides food that is a danger to public health or provides food that is contaminated for sale for human consumption or a person who contravenes a specific ministerial regulation adopted under section 54 of the Act shall be guilty of an offence and should be liable on summary conviction to a fine not exceeding £1,000 and in the case of a continuing offence to a further fine not exceeding £100 for each day on which the offence is continued or at the discretion of the court to imprisonment for any term not exceeding six months or to both fine and imprisonment.

The Health Act 1947 further provides the Minister with the ability to ensure that authorised officers may inspect food premises and take samples as they deem fit. Any authorised officer who has gained access to information by virtue of such inspections made in the enforcement of the Health Act or any regulations made under it shall not disclose such information unless it is necessary to do so for the enforcement of the regulations or the Act. Any authorised officer who contravenes these regulations in this way will be guilty of an offence equivalent to the standard fine/imprisonment as mentioned above. Section 61 of the Act introduces an offence whereby any person who wilfully obstructs the execution of regulations adopted under the 1947 Act, *e.g.* obstructing authorised officers to inspect buildings or to take samples may be prosecuted. In such a situation, a person will be liable on summary conviction to a fine not exceeding £1,000 and in the case of a continuing offence to a further fine not exceeding £100 for each day on which the offence is continued or at the discretion of the court to imprisonment for a term not exceeding six months or to both a fine and imprisonment.

Food Standards Act 1974

Mention should also be made of the Food Standards Act 1974 which provides the Minister for Agriculture and Fisheries, the Minister for Industry and Commerce or the Minister for Health with the power to make regulations providing for standards in relation to food and in particular to any of the following:

- names of foods;
- descriptions of foods;
- composition and quality of food;
- methods of manufacture and preparation of food;
- additives used in the manufacture and preparation of food;
- contaminants (including pesticide residues in food;
- hygiene in relation to food;
- time limits for consumption of food;
- packaging, labelling and presentation of food;
- transportation, storage and distribution of food;
- weights and measure for food; and
- matters consequential on or ancillary, incidental or supplementary to any of the foregoing matters.

Such regulations may provide that a person shall not import food intended for import, transport, store or sell food intended for human consumption under the name specified in the regulations if the food does not comply with these regulations. The Food Standards Act may also introduce regulations ensuring that a person shall not manufacture, prepare, import, transport, store or sell food if it does not comply with the regulations. Any person who contravenes a regulation made under the Food Standards Act shall be guilty of an offence. The Act goes on further to note that the minister must publish in a daily newspaper his intention to make any regulations under the Food Standards Act 1974. The Food Standards Act 1974, has only been utilised on one occasion since to introduce specific Irish food regulations in relation to potatoes. Obviously, since the majority of food law is now dependent on legislation from the European Union this particular Act has been somewhat under utilised but as it is still on the statute books it does provide the minister with considerable powers to enact specific Irish food law.

1.4 Miscellaneous UK Acts affecting the food sector

Weights and Measures Act 1985

This Act covers various statutory obligations concerning units and standards of measurement, weighing and measuring for trade, public weighing or measuring equipment and the regulation of transactions in goods.

Prevention of Damage by Pests Act 1949

This Act provides for the control of pests which are infesting food premises. Infestation is defined as "the presence of rats, mice, insects or mites in numbers or other conditions, which involve an immediate or potential risk of substantial loss or damage to food".

Trade Descriptions Act 1968

The Act specifies that any person who in the course of a trade or business:

(1) Applies a false trade description to any goods, or

(2) Supplies or offers to supply any goods to which a false trade description is applied

shall be guilty of an offence. The issue of a trade description is defined in section 2 of the Act and a false trade description is defined as one which is false to a "material degree". It is also specified that a trade description, which though not false is misleading shall also be deemed to be a false trade description under the Act and a person, can therefore be found guilty of an offence.

Animal Health Act 1981

This Act provides the Minister for Agriculture, Fisheries and Food with powers to make orders for the purpose of preventing the spread of disease. Section 10 specifically provides for orders for the purpose of preventing the introduction of diseases into Great Britain through the importation of animals and carcasses; carcasses of poultry and eggs and other things (whether animate or inanimate) by or by means of which it appears that any disease might be carried or transmitted.

Food and Environment Protection Act 1985

This Act covers three main areas:

(1) Powers to make emergency orders where an escaped substance may have contaminated food.

(2) Matters relating to dumping at sea.

(3) The control of pesticides.

If the escape of a substance is likely to create a hazard to human health through human consumption of food and the food may have been in an area of the United Kingdom, Ministers may make an emergency order under this Act. The order may prohibit certain specified things including preparation, processing and movement of food. The first such order under the Act was made in June 1986 to prohibit the movement or slaughter of sheep from parts of Cumbria and North Wales following the escape of radioactivity from Chernobyl, USSR.

1.5 **Conclusions**

The Food Safety Act 1990 was the outcome of two major government initiatives on food law in the late 1980s. The first of these was the publication of the White Paper "Food Safety – Protecting the Consumer" in July 1989. The second was the report produced by the Committee on the Microbiological Safety of Food, or better known as the "Richmond Committee" (named after its chairman). All of these documents issued recommendations as to how UK food safety policy might be amended in order to overcome the weak enforcement of UK food law at the time, which was having a major negative effect on consumer confidence in the food industry.

The most important aspects of the Food Safety Act 1990, which obtained Royal Assent on 29 June 1990, were the strengthening of enforcement powers including:

- the detention and seizure of food;
- the offence of supplying food that fails to comply with food safety requirements;
- powers to require registration of food premises;
- the issuing of improvement notices and the swift closure of premises if public health is threatened;
- the powers given to ministers to tackle potentially serious problems with emergency closure orders;
- tougher penalties and the modernisation of the system of statutory defences, especially the "due diligence" defence to cover all those operators in the food supply chain.

There is little doubt, therefore, that in 1990, this was a major reforming and radical piece of legislation. As the Food Safety Act imposed additional responsibilities on local authorities to enforce UK food law, the government granted an additional £30 million to local authorities to assist them in carrying out these new responsibilities.

However, by the end of the 1990s, food scares over salmonella in eggs, dioxin contamination in Belgium, the introduction of genetically modified foods and most importantly of all, the BSE crisis, prompted many commentators to maintain that UK food law was not adequately protecting the consumer. How could such a situation have occurred, following the adoption of the innovative Food Safety Act 1990? Despite all the recent difficulties in UK food safety law, the 1990 Act still plays the leading role in protecting consumers' health and providing a "level playing field" within which the food industry can operate. It is the foundation upon which UK food law is built.

Chapter 2

Product Liability and Product Safety

Summary

- EU Product Liability Directive 85/574/EEC
- Strict liability regime introduced into UK law
- Case law – "development risks" defence
- Extension of Product Liability Directive to include primary agricultural products
- EU Green Paper on Product Liability
- EU General Product Safety Directive 92/59/EEC
- General Product Safety Regulations 1994 – case law
- Product liability and general product safety law in Ireland.

2.1 Introduction

In the wake of the BSE crises, which undermined existing European Union food law it was decided to alter radically the prevailing regulatory regime for foodstuffs. The pre-BSE approach to food law was set out in the famous *Cassis de Dijon* case, which enshrined the principle of the free movement of foodstuffs throughout the internal market. In future, however, the EU has decided that enhancing food safety and consumer protection rather than the free movement of foodstuffs are to be the cornerstones of a new EU regulatory regime for foodstuffs. In order to meet the consumer's expectations regarding food safety, the European Commission has at its disposal a number of legal instruments concerned with compensation for victims of defective products, normally known as the Product Liability Directive (Council Directive 85/374/EEC) and the General Product Safety Directive (Council Directive 92/59/EEC). These legal instruments require all producers to place on the market only food products which are safe and the legislation makes them liable for repairing any such damage caused by any defective products.

2.2 **Product Liability Directive**

Council Directive 85/374/EEC[1] approximated Member States' laws on the producer's liability for damage caused by safety defects in his products. The Directive introduced a system of "liability without fault" whereby producers are liable for the damage caused by a defect in their product where the victim provides evidence of the existence of the damage, the defect and the causal relationship between defect and damage. Under the Directive liability is borne by any person placing a product on the market irrespective of whether he is:

- the manufacturer of a finished product;
- the producer of any raw material;
- the manufacturer of a component part;
- any person who by putting his name, trade mark or other distinguishing feature on the product presents himself as its producer;
- any person who imports a product into the European Union for sale, hire, leasing or any form of distribution;
- the supplier, where the producer or person furnishing the product cannot be identified.

Under Article 4 of the Directive "the injured person shall be required to prove the damage, the defect and the causal relationship between defect and damage". If several persons are deemed liable for the same damage, they are liable jointly and severally. A product is defective when it does not provide the safety which a person is entitled to expect, taking all circumstances into account, including the presentation of the product, the use to which it could reasonably be expected that the product would be put and the time when the product was put into circulation. It is also important to note that a product shall not be considered defective for the sole reason that a better product is subsequently put into circulation.

Under Article 7 the producer is provided with a number of defences allowing him to avoid liability under the Directive if he can prove that:

- he did not put the product into circulation;
- the defect which caused the damage did not exist at the time when the product was placed on the market or that the defect came into being afterwards;

1　OJ 1985 L210/29.

- the product was not manufactured in order to be placed on the market;
- the defect results from the product's compliance with mandatory regulations issued by the appropriate public authorities; or
- the state of scientific and technical knowledge when the product was put into circulation did not enable the defect to be discovered.

The Product Liability Directive stipulates that an action must be brought within three years from the date on which the defect occurs and that all possible actions are extinguished 10 years after the product has been placed on the market.

Under Article 9 there is a threshold of 500 Euros in relation to damage caused by a defective product and which is duly proven. In that case, the injured person is not fully compensated therefore such a person has to bear the first 500 Euros in any court action. Related to this stipulation, Article 16 permits Member States to decide that a producer's total liability for damage resulting from death or personal injury caused by a defective product can be limited to an amount of Euro 70 million. The Product Liability Directive was implemented into UK law by Part I of the Consumer Protection Act 1987 and into Irish law by the Liability for Defective Products Act 1991. Both Acts follow the provisions of the Product Liability Directive especially in relation to the development risks/scientific knowledge defence and also Article 2 of the Directive which provided Member States the possibility of a derogation to exclude primary agricultural products from the ambit of the legislation if they so wish. Before the European Commission proposed to amend the Product Liability Directive in 1998/99 this derogation had not been utilised by Greece, Luxembourg, Sweden and Finland.

Case law on the Consumer Protection Act 1987

Although there have been a number of UK cases involving sections of the Consumer Protection Act 1987, none of these deals specifically with food law issues and therefore it is not necessary to discuss them in this work. There has, however, been one important European Court case involving the UK implementation of the Product Liability Directive by the Consumer Protection Act 1987. The European Commission took an Article 226 (old Article 169) action against the United Kingdom stating that they had failed to implement correctly Article 7(e) of the Directive:

Commission v *United Kingdom.*[2] Article 7(e) states that the producer can have a defence against prosecution under the legislation if he can prove:

> "(e) that the state of scientific and technical knowledge at the time when he put the product into circulation was not such as to enable the existence of the defect to be discovered."

The Commission argued that under section 4(1)(e) of the Consumer Protection Act 1987, the UK legislature had broadened the defence under Article 7(e) to a considerable degree and converted the strict liability imposed by Article 1 of the Directive into mere liability for negligence. The UK government argued that section 4(1)(e) set out the same test as Article 7(e). The Court ruled that the Commission had failed in its action and concluded that the UK legislation adequately implemented the "development risks" defence provided for in the Product Liability Directive.

Amendment of Product Liability Directive

In May 1997 in response to the BSE crises, the European Commission published a Green Paper on food law as a means of launching a public debate on the likely changes envisaged for the future EU regulatory regime for foodstuffs. The Green Paper discussed issues such as food labelling, food hygiene, food quality, as well as the EU's important international obligations in the area of food law. Amongst a number of likely future legislative proposals, the European Commission noted its interest in amending the Product Liability Directive in order to being primary agricultural products within the ambit of the Directive. The Commission considered that extending the Directive's scope in this way would not dispense with the need for appropriate rules concerning product safety and efficient official control mechanisms in the Member States, therefore the extension of the Directive to include primary agricultural products in that case would merely constitute a complimentary measure.

In preparing a legislative proposal the Commission took the following factors into account:

- ▪ There was increasing public expectation of greater protection of public health and the Commission believed that this

---

2 Case C-300/95, *The Times* Law Report, 23 June 1997.

proposal would provide consumers with enhanced confidence in the safety of foodstuffs in general.

- A number of Member States, including Greece, Luxembourg, Finland and Sweden already had their own national laws on liability of agricultural producers, and therefore there was a need to harmonise product liability law throughout the European Union.

- The agricultural economies of those countries that already had national liability laws on primary agricultural products had suffered no major irreversible effects and therefore it was believed that extending the Product Liability Directive in this way would not impact negatively on the agricultural sector.

- Since the adoption of the Product Liability Directive there had been uncertainty as to the scope of the Directive in relation to products which had undergone "processing" since the Product Liability Directive covers products of the soil, stock farming and fisheries which have undergone "initial processing". This concept constitutes the dividing line between excluded agricultural products, *i.e.* primary products and those which the Directive covers, *i.e.* processed products. It is obvious that agricultural production does not escape the effects of industrialisation (*e.g.* the use of preserving techniques, deep freezing, etc which might involve risks), but the extent of such industrialisation is not apparent. The question is at what point does the use of a technique on a primary agricultural product involve "initial processing"? The Commission therefore considered that any uncertainty regarding this concept might discourage consumers from pursuing claims and therefore including primary agricultural products within the ambit of the Directive would put an end to these uncertainties.

Limitation periods

In October 1997, therefore, the European Commission published a proposal to the effect that it advocated the inclusion of primary agricultural products within the ambit of the Product Liability Directive. This proposal was finally adopted into EU law on 10 May 1997 as EP and Council Directive 1999/34/EEC.[3] The United Kingdom and Ireland

---

3   OJ 1999 L141/20.

will have until 4 December 2000 to implement this Directive into national law. As the Product Liability Directive stipulated (and implementing UK and Irish legislation followed this provision) an action must be brought within three years from the date on which the defect occurs and that all possible actions are extinguished 10 years after the product has been placed on the market. It is important to note that the amending Directive does not have retroactive effect, therefore actions can only be sought against primary agricultural products placed on the market after 4 December 2000 as mentioned previously. The liability under the Directive is strict but there are a number of defences available to the producer. The most important of these is the development risks defence whereby the producer is not held liable for the defect in a product if it was not possible to discover the existence of a defect due to the state of scientific and technical knowledge when the product was put into circulation. This defence will continue to exist in the case of primary agricultural products and could prove to be of great importance.

Particularly in the case of foodstuffs and pharmaceuticals, it is only after a fairly lengthy period of time that illnesses or damage caused by defective products appear and the causes of the damage can be established. In the case of BSE, it was only possible to prove the connection between defective products and the subsequent illnesses after a period exceeding 10 years. The European Parliament during the passage of the proposal amending the scope of the Product Liability Directive sought to have the limitation period extended in such cases, but that amendment was rejected by the European Commission and Member States. If genetically modified seeds/plants are found to be defective in future, the 10-year limitation period allied to the "development risks" defence may prove to deny consumers the right of redress for damage caused by such products. This is an outcome that the EU did not envisage when preparing to extend the Product Liability Directive to primary agricultural products. It will be interesting to see whether Member States introduce their own legislation covering liability issues in relation to genetically modified seeds/plants.

Strict liability regime

The liability without fault ("strict liability") regime established by the Product Liability Directive means that in future an agricultural producer would not be able to escape liability by arguing that he has complied with existing regulations and taken all the necessary precautions. Some representatives of the agricultural sector consider that, when combined with the existing system of liability with fault, the outcome might lead to farmers bearing excessive "liability". Many agricultural producers, knowing that they are generally the first link in the food production chain, fear that legal proceedings will be pursued right back to them, triggering their liability without fault. In the knowledge that agricultural production is largely at the mercy of phenomena over which the producer has no influence (*e.g.* weather, pollution, natural disasters and accidents), the agricultural producer may legitimately wonder why liability should be borne by him. It is a major oversight that the extension of the scope of the Product Liability Directive does not include a clause which suspends its application in the case of damage resulting from such cases of *force majeure*.

Since consumers are paying ever more attention to the safety of food products, there is bound to be an increase in the number of legal proceedings seeking redress against agricultural producers. The growth in legal activity may result in higher insurance premiums for agricultural producers. Recent events like dioxin contamination in Belgium point to the necessity of having such a system of liability in place so that consumers may have the possibility of seeking redress against a minority of unscrupulous producers concerned simply with quick profits rather than food safety and consumer health. The European Union is hopeful therefore that the existence of a strict liability scheme to cover primary agricultural products will act as an incentive to the placing of the safe, healthy, high quality food products on the market. In that case it represents one element of the EU's planned overhaul of its regulatory regime for foodstuffs, which aims to enhance food safety rather than the free movement of goods in order to further protect consumer health.

2.3 Green Paper on Product Liability

In July 1999 the European Commission published a Green Paper on Product Liability[4] which it hopes will begin a process of consultation between all relevant parties in order to examine how the rules on liability in respect of defective products are actually applied and to assess their impact on the operation of the internal market, consumer protection and the competitiveness of European business.

---

4   Green Paper on Liability for Defective Products, COM (99)396, 28 Sept 1999.

The results of the Green Paper will serve to prepare the second report on the application of the Product Liability Directive, which is planned to be published at the end of 2000. The Green Paper aims to obtain practical information which will enable the European Commission to make an in-depth analysis of how the Directive is actually being applied in a different context to the one when it was adopted, particularly by virtue of the new impetus given to the policy for the protection of health and safety of consumers following the BSE crises. The Commission's Green Paper aims to enable it to obtain practical and factual information particularly from consumers and industry as to whether the Product Liability Directive and its amendment to include primary agricultural products is achieving its objectives and also to gauge reactions as to a possible revision regarding the most sensitive points of this legislation. The Commission has specified a number of issues which it feels may be open to revision in the Directive. These issues include:

- detailed arrangements for implementing the burden of proof imposed on victims;
- existence of financial limits Euros 500 and their justification;
- 10-year deadline and the effects of any change;
- lack of any obligation on producers to obtain insurance;
- dissemination of information on cases arising from defective products;
- supplier's liability; and
- the type of goods and damage covered.

The Commission invited any concerned party to submit written observations on the questions contained in the Green Paper by December 1999. These observations will assist it in preparing its Second Report on the Application of the Product Liability Directive by the end of 2000. The area of product liability is therefore likely to continue to be in a state of flux for the next few years. Any changes the Commission proposes to the existing strict liability regime will have important implications for the food industry.

2.4 General Product Safety Directive

Council Directive 92/59/EEC[5] was intended to fill a number of gaps in EU consumer safety legislation. It achieves this by specifying that

---

5   OJ 1992 L228/24.

the products supplied to consumers must be safe. The provisions of the Directive apply insofar as there are no specific provisions in rules of Community law governing the safety of the products concerned. In particular where specific rules of Community law contain provisions imposing safety requirements on the products which they govern, the provisions of Articles 2-4 of the Product Liability Directive shall not apply to those products. For the purposes of the Directive "**a product**"shall mean any product intended for consumers or likely to be used by consumers and supplied to them in the course of a commercial activity.

Safe products

A "**safe product**" shall mean any product, which under normal or reasonably foreseeable conditions of use does not present any risk or only the minimal risk compatible with the product's use to consumers, taking into account the following points:

- the characteristics of the product including its composition, packaging, instructions for use;
- the effect on other products where it is reasonably foreseeable that it will be used with other products;
- the presentation of the product, the labelling, any instructions for its use and disposal and any other indication of information provided by the producer;
- the categories of consumers at serious risk when using the product, in particular children.

Dangerous products

A "**dangerous product**" under the Directive means any product which does not meet the definition of a safe product as outlined above. A "**producer**" under the Directive means the manufacturer of the product where he is established in the European Union, the manufacturer's representative when the manufacturer is not established in the EU and other professionals in the supply chain insofar as their activities may affect the safety properties of a product placed on the market. On the other hand, a distributor under the Directive means any professional in the supply chain whose activity does not affect the safety properties of a product.

Article 3 of the Directive states "producers shall be obliged to place only safe products on the market". Within the limits of their respective activities it is understood that producers shall provide consumers with relevant information to enable them to assess the risks inherent in a product through its normal or reasonably foreseeable period of use, where such risks are not immediately obvious, without adequate warnings. Provision of such warnings does not, however, exempt any person from compliance with requirements to place safe products on the market as specified in Article 3 of the Directive.

A product is regarded as safe if it conforms to the specific EU provisions governing its safety. In the absence of such provisions the product must conform to the specific national health and safety rules applicable to that particular product before it can be marketed in another Member State. In the absence of EU or national rules, the conformity of a product will then fall under the General Products Safety Directive and it must be assessed having regard to:

- voluntary national standards giving effect to European standards;
- Community technical specifications;
- standards drawn up in the Member State in which the product is in circulation;
- codes of good practice in respect of health and safety in the sector concerned;
- the state of the existing scientific knowledge and technology.

Under the Directive Member States must establish administrative infrastructures to enable hazardous or dangerous products to be identified so as to ensure compliance with the "general safety" requirement in Article 3.

Under Article 6, Member States are provided with the necessary powers to adopt appropriate measures to impose particular restrictions on products, which they have found in their investigations to be unsafe. If a Member State takes such actions, under the Directive it must inform the European Commission, which is then required to issue an opinion on the appropriateness of the measures taken (Article 7).

Rapid alert system

In the case of a product which presents a serious and immediate risk extending beyond the territory of the Member State concerned, the Member State must notify the Commission that it has taken or is going to take emergency measures to restrict or prevent the marketing of that product. The Commission will check to see whether the product complies with the provisions of the Directive and will forward the information to the other Member States which, in turn, must immediately inform the Commission of any measures they adopt to deal with the problem.

This system of exchange of information is generally known as the "**rapid alert system**" which is available in particular in the case of food emergencies. The General Product Safety Directive points out that effective supervision of product safety requires the establishment at national and EU levels of a system of rapid exchange of information about specific products in emergencies. An Annex to the Directive sets out the detailed procedures whereby the Commission will inform other Member States about the possible threat to consumer health from an unsafe food product. In 1999, for example, the rapid alert system came into play during the dioxin contamination scandal in Belgium.

If Member States do not forward information on dangerous food products, they risk allowing other Member States to take very stringent unilateral measures beyond the actual risk involved. In the case of unsafe food products from abroad a safeguard clause allows for the closing of borders. In certain circumstances and in particular where the Member States differ on the measures to be taken with regard to food emergencies and where the specific Community procedures prove inadequate to deal with the risk, the Directive provides for a Community procedure for the adoption of emergency measure. Under this procedure, the Commission, assisted by a Committee on Product Safety Emergencies, may demand that Member States take identical measures within a certain time limit with regard to the allegedly dangerous food product. In the case of dioxin contamination in Belgium, a number of very stringent measures were adopted in the form of different Commission Decisions. All Member States were then duty-bound to implement these measures immediately in order to reduce the impact of this food emergency on consumer health.

2.5 General Product Safety Regulations 1994

The General Product Safety Regulations 1994 (SI 1994 No 2328) came into force on 3 October 1994 and implement the safety provisions of the General Product Safety Directive. The core requirements of the 1994 Regulations closely follow the general safety requirement in

Part 2 (s 10) of the Consumer Protection Act 1987 which had been in force in the United Kingdom since October 1987. The Regulations largely replace section 10 of the 1987 Act. However, section 10 will remain in force for use in certain very limited circumstances. The General Product Safety Directive and hence the 1994 Regulations do not allow for products to be CE marked. **CE marking** is reserved for products covered by separate EU provisions whereby the marking is a declaration that the product in question satisfies the particular requirements of that law and creates a presumption (which is rebuttable) that the product is entitled to free circulation within the European Union. It should be noted that CE marking a product which does not require to be so marked is likely to constitute an offence under section 1 of the Trade Descriptions Act 1968.

The Consumer Safety Unit of the Department of Trade and Industry has overall policy responsibility for the Regulations, but the departmental sectoral responsibilities for safety matters remain unchanged, *e.g.* where a question of food safety arises, the responsible departments will continue to be the Ministry of Agriculture, Fisheries and Food (MAFF), the Department of Health, the Scottish Office Agriculture and Fisheries Department and the Northern Ireland Department of Health and Social Services. Likewise, the Medicines Control Agency and the Veterinary Medicines Directorate take the lead on licensed medicines for human and veterinary use respectively.

The Regulations apply to new and second-hand consumer products except products that are covered by specific European safety legislation. They apply to a wide range of products but not to service activities. "Products" within the meaning of the Regulations can best be described as goods which are supplied to consumers for their private use (reg 2(1)). Products covered include but are not restricted to:

- clothing,
- medicines,
- d.i.y. tools and equipment,
- food and drink,
- household goods,
- nursery goods,
- chemicals and pesticides.

Although the product coverage is wide, most will already be subject to national safety requirements. Only such goods supplied in the course of a commercial activity, which includes a business and a trade, are covered. The Regulations do not apply to products used solely in the workplace. Products which are used in the provision of a service, even if they are used for or by consumers are also outside the scope of the Regulations, *e.g.* escalators in a shopping centre or shopping trolleys provided by a supermarket for use by consumers. Health and safety at work legislation controls the safety of such products.

In keeping with the General Product Safety Directive, the Regulations make no provision for exemptions as such but they do not apply in cases which a product is subject to specific EU legislation. Where a product is subject to specific EU legislation, which is comprehensive in terms of safety, *i.e.* covers all aspects of the safety of a product (reg 3(c)) the Regulations are completely disapplied. However, where specific EU legislation covers only certain safety aspects of the product, the Regulations will apply to the remaining aspects of that product (reg 4). When assessing whether a Directive is comprehensive in terms of safety, both the intent of the particular Directive and its coverage must be considered. Products for export to a country outside the European Union where the goods are exported directly by the UK manufacturer (*i.e.* the products are not placed on the EU market) and are not in circulation before being exported are exempted from the General Product Safety Regulations.

The Regulations place a general duty on all suppliers of consumer goods to supply safe products. Safety takes into account factors such as the product's characteristics, instructions and warnings, and the categories of consumer at risk when using the product, particularly children. Any relevant British or European standards may be taken into account in assessing the safety of a product. Products must provide the safety which consumers may reasonably expect and compliance with the safety standard will not therefore guarantee that a product satisfies the general safety requirements, unless that standard adequately covers all aspects of the safety of the product concerned. The Regulations apply to all persons in the business supply chain who are established in the United Kingdom and supply consumer goods in the United Kingdom – whether the goods are intended for consumption in the United Kingdom or another Member State of the European Union. In this context, suppliers include manufacturers, importers, wholesalers and retailers.

Penalties and offences

It is an offence (regs 12 and 13) for:

"A producer to place a product on the market unless it is safe or offer or agree to place on the market or expose or possess for placing on the market a dangerous product;

A distributor to supply a product unless it is safe or offer or agree to supply or expose or possess or supply a dangerous product."

Supplying an unsafe product can result in a fine of up to £5,000 for each offence and/or a term of imprisonment for up to three months.

In cases where the Regulations apply, they provide a "due diligence defence" (reg 14(1)) which allows a producer or distributor to argue that he took all reasonable steps and exercised all due diligence to avoid committing the offence. Where a producer or distributor argues the defence that the commission of the alleged offence was due to the act or default of another person or due to reliance on information given him by another person (reg 14(2)) he must notify this defence to the prosecution not less than seven days before the date of hearing the case. It should be noted that this is not an automatic defence and will require justification to the court. Where a person (the principal offender) has committed an offence under the regulation and this was due to the act or default of another person proceedings may be brought against that other person whether or not proceedings are also brought against the principal offender (reg 15(1)).

If it can be shown that an offence has been committed with the consent or connivance or is attributable to any neglect on the part of any director, manager, secretary or other similar officer of the body corporate, such persons (in addition to the body corporate) may also be proceeded against (reg 15(2) to (4)). The Regulations introduce a time limit for bringing prosecutions and thereby require that any prosecution for an offence under the Regulations must be brought within 12 months from the commission of that offence (reg 16).

Enforcement

The responsibility for day-to-day enforcement of the regulations rests with local trading standards authorities in England, Wales and Scotland and in Northern Ireland with the District Councils. Where the safety of a product is covered by national legislation, *i.e.* an Act or specific Regulations, that legislation will continue to provide the means for assessing safety and accordingly should be considered to be an appropriate vehicle for any necessary enforcement of action.

In the example of food, the Food Safety Act 1990 (and the Regulations made thereunder) is recognised as dealing with significant safety issues relating to food. In the case of food safety, the 1990 Act would prevail and food satisfying the requirements of the 1990 Act should be presumed to satisfy the requirements of the General Product Safety Regulations. Relevant enforcement action should be taken under the provisions of the 1990 Act and businesses should not be placed under double jeopardy by parallel action for the same matter also being taken under the General Product Safety Regulations 1994.

The Local Authorities Co-ordinating Body on Food and Trading Standards (LACOTS) promotes the "home authority" principle, which applies to all food hygiene, food safety and trading standards matters. Under this principle, the local authority where an enterprise is established accepts the primary responsibility for offering advice and preventative guidance on a regular basis on safety and other related matters to the enterprise. Other local authorities are expected to liaise with the relevant home authority on any safety matters arising from the product supplied by that business. Businesses are therefore encouraged to make contact with and seek advice on any particular matter from their home local authority. The home authority principle is aimed at promoting uniformity of approach to trading standards matters, reducing duplication and assisting businesses to comply with the law.

Case law on General Product Safety Regulations 1994

The courts have held that where a person is proceeded against by virtue of regulation 15(1) of the General Product Safety Regulations 1994, the 12-month limit in regulation 16 applies: *R v Thames Magistrates' Court ex parte Academy International plc.*[6] The court ruled that a party accused of an offence relating to dealing in dangerous products under regulation 13 of the 1994 Regulations could raise a defence of due diligence under regulation 14(1). Where such a defence involved an allegation that the offence was due to the act or default of another party, regulation 15 provided that the second party could be proceeded against. Regulation 16(1) applied a 12-month time limit to prosecutions for offences contrary to regulation 13. The defendants, who had been prosecuted as secondary parties under the regulation 15 procedure, argued that in such circumstances the usual time limit of six months applied. The court ruled that the 12-month limit applied.

---

6   *The Times* Law Report, 23 June 1999.

In another case, *Coventry City Council* v *Padgett Brothers Ltd*[7], the courts ruled that the commission of an offence contrary to regulation 15(1) of the 1994 Regulations "due to the act or default committed by some other person" was made out against the importer of the goods when the retailer had been convicted of the offence of supplying dangerous goods contrary to regulation 13(b) of the same Regulations.

2.6 **Product liability and general product safety legislation in Ireland**

The Product Liability Directive has been implemented by the Liability for Defective Products Act 1991, which follows the EU Directive in most matters, although like the United Kingdom, Ireland did seek a derogation to avoid primary agricultural products coming within the scope of the legislation. The General Product Safety Directive has been implemented by the General Product Safety Regulations 1997 (SI 1997 No 197). These Regulations make it an offence to place unsafe products on the market. They specify the duties of producers and distributors in relation to placing safe products on the market. The Irish Director of Consumer Affairs is given the authority to ensure that only safe products as specified in the regulations are placed on the market.

A number of offences are listed in the Regulations whereby if the producer fails to place safe products on the market or fails to follow directions established by the Director of Consumer Affairs whether in terms of safety of products, labelling, consumer information etc the person shall be guilty of an offence and can be liable on summary conviction to a fine not exceeding £1,500 or to imprisonment for a term not exceeding three months or to both. Where a person after conviction continues to contravene the provision concerned he can be liable on summary conviction to a fine not exceeding £250 for every day in which the contravention continued.

2.7 **Conclusions**

Although according to the European Commission, there has been no significant increase in product liability/safety claims, and insurance claims are generally unaffected by this legislation, there is certainly a

---

7   *The Times* Law Report, 24 Feb 1998.

perception that there is an increasing demand for legal advice on product liability and product safety issues. Forward-thinking businesses are now working with external legal advisers on "stewarding" their products through specific life cycles, ensuring that the potential for defects is kept to a minimum. This is particularly the case in the food industry, especially amongst companies with well-known brand name food products. Product liability legislation is not always driven by science and reason alone. Extending the Product Liability Directive to cover primary agricultural products, was very much in response to the BSE crisis. Although the European Commission believes that it will not increase the number of legal proceedings taken under the Directive, there are many who disagree. Some lawyers think that insurance premiums for the farming community will increase at a time when they are facing considerable difficulties over the reform of the CAP and increased world competition in prices for agricultural products.

Of greater concern for the food industry, are the proposals outlined in the Commission's Green Paper to reform the strict liability regime provided for in the Product Liability Directive. The proposals, which could cause most difficulty to the food industry, would be the reversal of the burden of proof and the removal of the "developments risk" defence. Reversing the burden of proof could encourage spurious claims. The Commission is also undertaking a review of the General Product Safety Directive, in particular it is looking at amending the definition of the term "safety". This may be with an eye to adopting the "precautionary principle" by which a product is assumed to be unsafe until there is sufficient evidence to show that it is safe. This principle is already established in the environmental context, but if used in relation to food products, it could create serious problems for producers and suppliers of food products.

Another important factor the food industry would need to take into account in relation to product liability/safety issues is to be prepared to face actual court actions in a number of foreign jurisdictions. For example, the Coca-Cola cans which were contaminated in Belgium, many batches of these products which were destined for the French and Belgian markets were found far beyond their main sale areas months after the recall notice was issued.

The whole legislative framework for product liability and product safety is presently in flux. The Product Liability Directive has been extended to primary agricultural products, which will come into operation in all Member States in December 2000. The product liability regime is also under discussion in the Commission's Green Paper, as is the General Product Safety Directive in the present "review process" by the Commission. Many believe that GM foods and biotechnology products must be covered by separate liability regimes in the future, as the Product Liability/General Product Safety Directives are inadequate for such innovative products. In that case, this area of food law is certain to develop in the coming years.

Chapter 3

· **Food Hygiene** ·

Summary

- General Food Hygiene Regulations 1995
- Hazard Analysis and Critical Control points (HACCP) principals introduced
- Temperature Control Regulations 1995
- Egg Products Regulations 1993
- Imported Food Regulations 1997
- Food hygiene law in Ireland

3.1 **Introduction**

The principal food legislation relating to food hygiene is based either on the vertical hygiene Directives or the General Food Hygiene Directive both emanating from Brussels. The European Commission clearly stated in its Green Paper on Food Law (May 1997) that the field of food hygiene raises some of the most difficult questions in relation to possible future reforms of EU food legislation. At the same time, health risks related to poor hygiene conditions in the production and preparation of foodstuffs are increasingly seen by consumers and food experts as a far greater risk to human health than other food safety issues, such as additives and genetically modified foods. Certainly, the risk to human health from poor hygiene conditions is more "immediate" than that from additives and genetically modified foods and every week the media reports on instances of salmonella poisoning caused by poor hygiene conditions whether in the production, preparation or retailing of foodstuffs.

The present EU food hygiene legislation is a "patchwork" of different pieces of legislation aimed at legislating for a specific product group, *e.g.* egg products, fresh meat, poultry meat etc. At the same time these vertical hygiene Directives co-exist with the General Food Hygiene Directive 93/43/EEC.[1] The co-existence of these various

---

1   OJ 1993 L175/1.

Directives has resulted in numerous criticisms of inconsistency and incoherence. In particular, Article 1(2) of the General Food Hygiene Directive requires the European Commission to establish definitively the relationship between these different pieces of legislation. The Commission is presently working on a draft proposal for a European Parliament and Council Regulation on the hygiene of foodstuffs, which aims to recast the vertical hygiene Directives in one overall piece of EU legislation. It may take many years before this proposal is finally adopted into EU law. In order to see how EU food hygiene law has been implemented into UK law, the implementation of the General Food Hygiene Directive and one of the vertical hygiene Directives, egg products, will be discussed in detail.

3.2 **Food Safety (General Food Hygiene) Regulations 1995**

These Regulations (SI 1995 No 1763) implement Council Directive 93/43/EEC on the hygiene of foodstuffs, except for the requirements of paragraphs 4 and 5 of chapter 9 of the Annex to that Directive (which relate to temperature controls); these have been implemented by a separate regulation. A separate regulation also implements the General Hygiene Directive requirements in relation to imports which are likely to pose a serious risk to human health and which originate from countries that are not part of the European Union. It is clearly specified in regulation 3 of the 1995 Regulations that they do not apply to the various UK regulations which have implemented the separate vertical hygiene Directives, *e.g.* the 1995 Regulations do not apply to the Egg Products Regulations 1993 (SI 1993 No 1520) or the Fresh Meat Regulations 1995 (SI 1995 No 539) etc.

HAACP principals

The General Food Hygiene Directive introduced what are known as Hazard Analysis and Critical Control Points (HACCP) principals which must be developed and introduced at all stages of the food chain. The 1995 Regulations include the HACCP principals in regulation 4, by specifying that a proprietor of a food business shall ensure that any of the following operations, *e.g.* the preparation, processing, manufacturing, packaging, storing, transportation, distribution, handling and offering for sale or supply of food are carried out in a "hygienic way". In order to do this, a proprietor of a food business shall ensure that the requirements of various Schedules within the Regulations are adhered to.

Hygiene rules

Chapter 1 of this Schedule specifies various rules of hygiene that must be followed, noting that food premises must be kept clean and maintained in good repair and condition. The layout, design, construction and size of food premises should permit adequate cleaning, so as to protect against the accumulation of any dirt and the shedding of particles into food. Good food hygiene practices should be followed so as to protect against cross-contamination and, where necessary, suitable temperature controls for the hygienic processing and storage of products should be adhered to. Any food premises should have an adequate number of wash-hand basins suitably located and designated for cleaning purposes. An adequate number of flush lavatories must be available and connected to an effective drainage system. Wash basins for cleaning hands must be provided with both hot and cold water, as were materials for cleaning and for drying hands in a hygienic fashion. The ventilation in the food premises must be of a suitable nature and there should be a sufficient means of airflow throughout the premises. Food premises must also have adequate natural and/or artificial lighting and adequate changing facilities for personnel must be provided where necessary.

The proprietor of a food business must also ensure that various other specific requirements in other chapters of the Schedule are followed, whether with regard to floor surfaces, wall surfaces, ceilings and overhead fixtures, windows and other openings, doors and any working surfaces that would be in contact with food. The issue of the cleaning and disinfecting of both work-tools and equipment is also dealt with in these chapters. Finally, there are specific requirements for moveable and/or temporary premises (such as marquees, market stalls, mobile sales vehicles) or such premises used primarily as a private dwellinghouse or premises used occasionally for catering purposes as well as vending machines.

Food safety procedures

In all these cases, a proprietor of a food business should identify any step in the activities of the food business which is critical to maintaining food safety and he must ensure that adequate safety procedures are identified, implemented, maintained and reviewed on the basis of what are known as the HACCP principals. Regulation 4(3) outlines these principals as follows:

(a) analysis of the potential food hazards in a food business operation;

(b) identification of the points in those operations where food hazards may occur;

(c) deciding which of the points identified are critical to ensuring food safety ("critical points");

(d) identification and implementation of effective control of monitoring procedures at those critical points; and

(e) review of the analysis of food hazards, the critical points and the control and monitoring procedures periodically and whenever the food business operations change.

Regulation 5 specifies that a person working in a food handling area who may be suffering from a disease or has a particular skin infection or a wound or is suffering from diarrhoea or another medical condition should make this information known to the proprietor. The proprietor will then ensure that that person is refused permission to work while under medical supervision. Regulation 6 specifies that any person who contravenes either regulation 4 or 5 of the Regulations will be guilty of an offence and can be liable:

- on summary conviction to a fine not exceeding the statutory minimum; or
- on conviction on indictment to a fine or imprisonment for a term not exceeding two years or both.

It is important to note that through regulation 7 the Regulations ensure that the various sections of the Food Safety Act 1990 apply to these Regulations and that the due diligence defence is therefore available to a food proprietor who may be prosecuted under this Act. Finally, regulation 8 specifies that each local food authority shall enforce and execute the Regulations within its own area. In order to do this they shall ensure that food premises are inspected on a regular basis and that such inspections will include a general assessment of the potential food safety hazards associated with the food business being inspected.

3.3 **Food Safety (Temperature Control) Regulations 1995**

In conjunction with the General Food Hygiene Regulations 1995, the UK authorities introduced these Regulations (SI 1995 No 2000) which implement the specific aspects of the General Food Hygiene Directive relating to temperature control. As with the previous Regulations, regulation 3 of the Food Safety (Temperature Control) Regulations notes that they do not extend to the various UK regulations which have implemented the 11 vertical hygiene Directives, covering egg products or meat products etc.

Regulation 4 details the chill-holding requirements specifying that no person shall keep any food:

> "(a) which is likely to support the growth of pathogenic micro-organisms or the formation of toxins; and
>
> (b) with respect to which any commercial operation is being carried out at or in a food premises at a temperature above 8° centigrade."

Chill-holding requirements (reg 5) will not apply to food which:

- has been cooked or re-heated;
- is for service or on display for sale;
- needs to be kept hot in order to control growth of pathogenic micro organisms or the formation of toxins;
- for the duration of its shelf life may be kept at ambient temperatures with no risk to health;
- has been subjected to a process such as dehydration or canning intended to prevent the growth of pathogenic micro organisms;
- must be ripened or matured at ambient temperatures;
- raw food intended for further processing which includes cooking before human consumption;
- to which EU legislation on the marketing standards for either poultry or eggs applies.

Defences

Regulation 6 specifies that in any proceedings for an offence contravening a regulation 4 obligation it will be a defence for a person charged to prove that:

"(a) A food business responsible for the manufacturing, preparing or processing of the food has recommended that it be kept at or below a specified temperature between 8°c and ambient temperatures;

(b) The recommendation has, unless the defendant is that food business, been communicated to the defendant either by means of a label on the packaging of the food or by means of some other appropriate form of written instruction;

(c) The food was not kept by the defendant at a temperature above the specified temperature; and

(d) At the time of the commission of the alleged offence the specified shelf life had not been exceeded."

Regulation 6 includes the provision, which allows for the upward variation of the standard temperature in appropriate circumstances to act as a defence to any regulation 4 prosecution. Any such variation must, however, be based on a well-founded scientific assessment of the safety of the food at the new temperature. There are also defences which relate to the tolerance periods for which food may be held outside the temperature control (reg 7).

Regulation 8 contains a requirement that food which needs to be kept hot in order to control the growth of pathogenic micro organisms or the formation of toxins must be kept at a minimum temperature of 63°c. There are defences which allow for a downward variation of this minimum temperature in appropriate circumstances and for a two-hour tolerance period (reg 9). A new general temperature control requirement is introduced in regulation 10, which prohibits keeping perishable foodstuffs at temperatures which would result in a risk to health. Regulation 11 contains a further requirement in relation to the cooling of food.

Penalties and enforcement

Any person guilty of an offence against these Regulations (reg 17) shall be liable:

- on summary conviction, to a fine not exceeding the statutory minimum; and
- on conviction on indictment to a fine or imprisonment for a term not exceeding two years or both.

As with the previous Regulations various sections of the Food Safety Act 1990 also apply in relation to these Regulations; therefore, the most important of these (the due diligence defence) is available to anyone prosecuted under the 1995 Regulations. Finally, as with the previous Regulations, each food authority shall enforce and execute these Regulations within its own area by ensuring the food premises are inspected on a regular basis so as to ensure that each proprietor of a food business is acting in accordance with these Regulations or any other relevant guide to good hygiene practice which has been adopted in line with the EU General Food Hygiene Directive.

Amendments

Finally, the Food Safety (General Food Hygiene) Regulations 1995 have been amended by SI 1999 No 1360, which provides for a particular derogation from the Hygiene of Foodstuffs Directive, in order that the bulk transport of raw sugar by sea in receptacles, containers or tankers that are not used exclusively for the transport of foodstuffs will not fall under the provisions of the General Food Hygiene Regulations 1995.

3.4 **Egg Products Regulations 1993**

There are 11 vertical hygiene Directives which have already been implemented into UK law. Rather than deal with all 11 Regulations, the Egg Products Regulations 1993 (SI 1993 No 1520) will be examined in detail. These Regulations implement in part Council Directive 89/437/EEC on the hygiene and health problems affecting the production and the placing on the market of egg products. The Regulations also implement other aspects of EU law with regard to egg products but they are seen as the main piece of legislation regarding the health and hygiene provisions which would affect the production and placing on the market of egg products. In general, the Regulations make provision in the United Kingdom for the preparation and manufacture of egg products used in food intended for sale or for human consumption and they prohibit the manufacture of egg products other than in an approved establishment.

Definitions

Regulation 2 provides a number of definitions for various issues, such as "egg", "egg products", "handling" "supply", and "batch". It also specifies that various sections of the Food Safety Act 1990 will apply to these Regulations, in particular the defence of due diligence. The

Regulations (reg 3) introduce various obligations on those selling eggs for human consumption. There is a prohibition on the sale of any egg products which are a mixture of egg products obtained from more than one species. The egg products must be produced in line with the various Schedules of the Regulations which point out clearly that there should be an avoidance of all contamination during the production, handling and storage of egg products. The Schedules introduce various requirements concerning, for example:

- breakages of eggs;
- the temperatures at which eggs should be stored and packed;
- the pasturisation of whole egg or yoke;
- the determination of alpha amylase in whole egg or yoke;
- the heat treatment of albumen;
- micro biological criteria for each sample of egg products of each batch sent for sale;
- the test for salmonella and heat treated egg products and various other tests;
- bacteriological tests; and
- the storage and transport of egg products.

Egg products prepared in the United Kingdom must have been prepared in an approved establishment, approval being given by the local authority as laid down in Schedule 8 to the Regulations. Egg products from other EU countries must have been prepared in establishments which meet the equivalent EU standards. Regulation 4 specifies that any person applying heat treatment to egg products should keep accurate records and retain the records for a period of not less than two years and produce such records on request to the appropriate food authority.

As of 14 October 1993 no person is permitted to manufacture any egg products or apply heat treatment to any egg products for the purpose of sale for human consumption otherwise than in an establishment approved for the purposes of these regulations. Upon application by a proprietor the local food authority will issue an approval notice (reg 5) if it is satisfied that the egg products establishment complies with the various requirements in Schedules 1 and 8 to the Regulations. The appropriate Minister (reg 6) is given the powers to revoke an approval granted under regulation 5 in respect of any egg products establishment if, after inspection or inquiry, it is found that various requirements of the Regulations are no longer being complied with or there are breaches of such conditions.

The Regulations specify that no person should dispatch a container of non-pasteurised egg products from an approved establishment for treatment at another approved establishment unless various conditions are observed.

Penalties and enforcement

Any person contravening the Regulations shall be guilty of an offence and be liable:

- on summary conviction to a fine not exceeding £5,000; and
- on conviction on indictment to a fine or to imprisonment for a term not exceeding two years or both in relation to certain offences and to a fine in the case of other offences.

The local food authority is obliged and is responsible for the enforcement and execution of the provisions of these Regulations in its own area.

3.5 **Imported Food Regulations 1997**

These Regulations (SI 1997 No 2537) contain measures relating to the control of certain types of food imported into the United Kingdom, which are not in free circulation within the European Union. They replace the general provisions of the Imported Food Regulations 1994 (SI 1984 No 1918), by adding a new set of provisions which apply to all food other than specified exempt products of animal origin. Under regulation 3(1) a list of products of animal origin exempt from the scope of these Regulations is provided for in Schedule 1. The list contains animal origin products, which fall within the ambit of the 11 vertical hygiene Directives, dealing with hygiene and control measures for these products throughout the European Union.

Regulation 4 stipulates that no person shall import into the United Kingdom from a third country any food intended for sale for human consumption which:

- fails to comply with food safety requirements; or
- is unsound or unwholesome.

Enforcement

Under regulation 5 the various existing UK food authorities have the responsibility of enforcing these Regulations. Regulation 6 covers the procedures relating to the examination of imported food which comes within the scope of the Imported Food Regulations. It includes requirements that the importer must provide all such facilities as the authorised officer of a food authority may reasonably require for the examination of any batch, lot or consignment of food which is due to arrive in his food authority area. If an authorised food authority officer considers that a sample of any food should be procured, he may by notice in writing given to the importer require such a sample. Additionally, he can stipulate that the food product should not be removed from a specified place for a period not exceeding six days excluding Saturdays, Sundays and public holidays. Such a notice will be served on the importer and, if he feels aggrieved by the decision, he does have the facility to appeal against the notice to a magistrates' court. The magistrates' court may order that the notice be withdrawn or that a shorter period be fixed for examination of the food.

Regulation 7(1) to (4) contain notice procedures allowing for the re-export of food which fails to comply with the regulation 4 requirement to be safe for human consumption. Alternatively, it provides for the situation where the food authority will permit the import of such food but stipulates that due to food safety concerns, it will not be for human consumption. Again, the importer will be served a notice as to this effect stating the grounds which the authorised officer has for believing that the food fails to comply with specific food safety requirements. Any importer who is aggrieved by such a decision may within six days appeal against the decision to the magistrates' court. The magistrates' court may cancel or affirm the particular notice. Alternatively, in regulation 7(5) and (6) of the Imported Food Regulations there is provision for the food authority to seek to have such food destroyed in accordance with established Food Safety Act 1990 procedures (s 7 or 8 of the 1990 Act).

Penalties

Any person who contravenes any provisions of the Imported Food Regulations shall be guilty of an offence and can be held liable, on summary conviction to a fine not exceeding the statutory maximum and on conviction on indictment to a fine or imprisonment for a term not exceeding two years or both. Finally, regulation 9 allows for the application of certain provisions of the Food Safety Act 1990, including the section 21 defence of due diligence.

3.6 **Hygiene legislation in Ireland**

The General Hygiene of Foodstuffs Directive (Council Directive 93/43/EEC) has been implemented in Ireland by the European Communities Hygiene of Foodstuffs Regulations 1998 (SI 1998 No 86). Under regulations 15-28 obligations are laid down on proprietors of food businesses in relation to general food hygiene.

The proprietor of a food business shall ensure that any process in the activities of his food business which is critical to guaranteeing food safety is identified and he shall also ensure that adequate safety procedures are identified, implemented, maintained and reviewed on the basis of the principles used to develop the system of HACCP (Hazard Analysis and Critical Control Points). The HACCP principles are elaborated in regulation 3 of the General Food Hygiene Directive and the Irish Regulations generally follow the philosophy enshrined within the EU Directive.

Penalties and enforcement

The competent authorities in Ireland, which have been assigned to oversee the implementation and enforcement of the Hygiene Directive, are the various health boards. It is important to note that under regulation 13 it is stipulated that they do not apply to the various Regulations which implement the 11 vertical hygiene Directives covering areas such as wild game, rabbit meat, milk products, egg products, fresh poultry meat etc.

Under regulation 6 any person who contravenes any provisions of the Regulations shall be guilty of an offence and will be liable on summary conviction to a fine not exceeding £1,000 or at the discretion of the court to imprisonment for a term not exceeding six months. For the purposes of these Regulations, every contravention of a regulation shall be deemed a separate contravention and shall carry the same penalty as for a single contravention of any provision of these Regulations. There is also provision for a body corporate to be prosecuted under these Regulations.

Finally, proceedings for an offence under these Regulations must be instituted within 12 months from the date of the offence, or any time within 12 months from the date on which knowledge of the commission of the offence came to the attention of an authorised officer. There is also provision under regulation 9(5) for a health board, where appropriate, depending on the specific risks to human health, to withdraw and/or close all or part of a food business for an appropriate period of time in accordance with the provisions in the European Communities (Official Control of Foodstuffs) Regulations 1998 (SI 1998 No 85). Under regulation 16 of the Official Control of Foodstuffs Regulations a chief executive officer of a health board who has evidence that there is a grave and immediate danger to human health because a foodstuff is so diseased, contaminated, or otherwise unfit for human consumption, may apply to the justice of the district court for a closure order prohibiting the operation of the food business. The district court may then decide to grant or refuse to grant such a closure order. The chief executive officer of a health board must give written notice to a food proprietor of his intention to seek a closure order before the date of the court hearing. The proprietor of the food business who has received such notice can then apply to the district court for annulment of the closure order depending on the decision of the court.

Egg products

Directive 89/437/EEC on egg products has been implemented in Ireland by the European Communities (Egg Products) Regulations 1991 (SI 1991 No 293). These Regulations stipulate that a person shall not:

- produce an egg product for use as a foodstuff; or
- produce an egg product for use in the manufacture of a foodstuff; or
- package any such egg product unless they have obtained an approval for establishing and carrying out such activities by the Minister for Agriculture.

Where the Minister approves of an establishment he shall allocate a serial number to such a building and notify the owner in person.

Penalties and enforcement

Authorised officers have powers to enter such premises, to take samples and look at documentation to ensure that all the obligations under the

Regulations are being followed. Such an officer can also place an improvement notice on the establishment and the owner has 21 days to appeal such a notice to the district court. These Regulations also apply to imports of eggs and a person guilty of any offence under the Regulations is liable on summary conviction:

- to a fine not exceeding £1,000; or
- to imprisonment for a term not exceeding six months

3.7 **Conclusions**

Implementing strategies for ensuring the safety of the food supply chain is a key and fundamental activity for food processors, retailers and regulatory agencies. Recent events taking the example of the supply of contaminated meat in Scotland, which resulted in the Pennington Report, point to the fact that a breakdown in the food hygiene regulatory system can have immediate effects on the well being and health of consumers. These events have stimulated a national debate in the United Kingdom on the question of food safety. The government's response has been to establish the Food Standards Agency to enhance consumer confidence and ensure that the existing body of UK food hygiene law is adequately enacted and enforced. In line with the EU General Food Hygiene Directive, it is also promoting, the establishment of HACCP (Hazard Analysis and Critical Control Points) systems at every stage in the food chain.

The food hygiene situation is not helped by the "patchwork" of vertical hygiene Directives on specific products, *e.g.* eggs, milk, fresh meat, poultrymeat etc existing at EU level and implemented into UK law. These Directives are very descriptive, whereas the General Food Hygiene Directive is a more flexible instrument encouraging the use of HACCP systems at every stage of the food production and retailing chain. There is also a proposal to consolidate all the vertical hygiene Directives and encourage the use of HACCP systems for these products. This will bring some clarity to the situation but HACCP systems alone will not prevent food poisoning occurring in the future. HACCP is a good management tool, which should be used in conjunction with quality control systems, increased foodborne illness research by government agencies and regulatory compliance programmes. It is hoped that the Food Standards Agency will assist in this matter to ensure that food hygiene will no longer be the cause of the majority of the United Kingdom's food safety problems.

Chapter 4

· **Food Labelling** ·

Summary

- Food Labelling Regulations 1996
- Compulsory labelling – name – list of ingredients – net quantity – date of minimum durability – storage conditions – manufacturers name – particulars of origin
- Amendments to Food Labelling Regulations 1996
- Case law
- Specific labelling for food additives
- Food labelling law in Ireland

4.1 **Introduction**

Food labelling has grown in importance over the years and plays an important role in ensuring that the consumer has adequate protection when purchasing food products. Food labelling fulfils three essential requirements:

(1) Product identification.

(2) Consumer information.

(3) Product marketing.

To fulfil these requirements, clearly recognisable, legible, simple, understandable, interesting and informative labelling is needed which will not mislead the consumer. More and more consumers are demanding additional information on food labels relating to food allergies, presence of GMOs etc. Most Member States had adopted some form of legislation on food labelling prior to the EU drawing-up a framework directive on food labelling in 1979. This legislation has now become the core legal text defining food manufacturers' obligations in relation to the question of labelling their products.

4.2 **Food Labelling Regulations 1996**

In the United Kingdom the main framework Directive on food labelling, 79/112/EEC[1], on the approximation of the laws of Member States relating to the labelling, presentation and advertising of food stuffs and the various amendments of the 1979 Directive have been consolidated into a single text: the Food Labelling Regulations 1996 (SI 1996 No 1499). These Regulations apply to food that is ready for delivery to the ultimate consumer or to a catering establishment. The Regulations provide for a number of food products to be exempted from their ambit. These foodstuffs include:

- any specified sugar product;
- cocoa or chocolate products;
- honey;
- condensed or dried milk;
- coffee, coffee mixtures, coffee extract products, chicory extract products or other such designated products;
- hen eggs and various other types of eggs;
- spreadable fats;
- wines or grape musts, sparkling wines, liquor wines, semi-sparkling wines;
- spirit drinks;
- fresh fruit and vegetables;
- preserved sardines and preserved tuna; and
- bonito.

Under regulation 5 all food to which the Regulations apply should be marked or labelled with:

- the name of the food;
- a list of ingredients;
- the appropriate durability indication;
- any special storage conditions or conditions or use;
- the name and address of the manufacturer or packer or a seller established within the European Union;
- particulars of the place of origin if necessary to avoid misleading the purchaser to a material degree; and
- instructions for use if necessary.

---

1   OJ 1979 L33/1.

Name of food

If a name prescribed by law exists, it should be used and it may be qualified by other words which make it more precise. Schedule 1 specifies a number of obligations on food manufacturers/retailers regarding the **name** to be used when labelling food and food products. The Schedule lists a number of names for fish, which should be used when referring to various species of fish. If there is no name prescribed by law, a customary name may be used. Where no name prescribed by law nor a customary name exists, a name sufficiently precise to inform a purchaser of the true nature of the food and to enable the food to be distinguished from products with which it could be confused should be used. If necessary, this wording shall include a description of its use. A trade mark, brand name or fancy name should not be substituted for the name of a food. If a food product is powdered, dried, freeze dried, frozen, concentrated or smoked and the omission of such an indication could mislead the purchaser, the name of the food must include or be accompanied by the indication that the food product has undergone such treatment. Finally, meat that has been treated with proteolytic enzymes shall be accompanied by the word "tenderised" and food, which has been irradiated, shall include or be accompanied by the word "irradiated" or "treated with ionising radiation".

List of ingredients

The list of ingredients must be preceded by an appropriate heading, which consists of or includes the word "ingredients". **Ingredients** should be listed by weight in descending order, determined as at the time of their use in the preparation of the food. Exceptions are made in the case of water and volatile products, which are added as ingredients to a food, ingredients that are used in a food in concentrated or dehydrated form, which will be reconstituted during preparation of the food. If a product consists of mixed fruit, nuts, vegetables, spices or herbs and no particular one of these ingredients predominates significantly by weight, the ingredients may be listed in any order, provided that for foods consisting entirely of such a mixture the heading includes "invariable proportion".

The name of an ingredient shall be the name, which would be used if the ingredient were sold as a food, or the generic name as listed in

Schedule 3 to the Regulations. To take an example: the generic name for fat, as an ingredient would refer to any refined fat. When indicating "fat" on the ingredients list the term should include:

- "animal" or "vegetable" as appropriate; or
- a specific animal or vegetable or if appropriate "hydrogenated" along with the word fat in any list of ingredients.

Conditions are also laid down for the use of the word "flavouring" and "natural". An additive shall be listed by either the principal function it serves, *e.g.* colour, preservative, followed by its name and/or serial number (E-type number) or where the function of the additive is not given its entire name should be specified in the ingredients list. The following foods need not be marked or labelled with a list of ingredients:

- carbonated water;
- vinegar;
- cheese, butter, fermented milk and fermented cream;
- flour; and
- any drink with an alcoholic strength by volume of more than 1.2%.

Appropriate durability indication

Under regulation 20 the minimum durability of food shall be indicated by the words "best before" followed by the date up to and including which the food can reasonably be expected to retain its inherent properties if properly stored as specified. The term "use by" may be substituted for the "best before" format. Foods which are exempted from stating an appropriate durability indication include:

- wine, liquor wine, sparkling wine and any similar drink obtained from fruit other than grapes; any drink with an alcoholic strength by volume of 10% or more;
- any soft drink, fruit juice or fruit nectar;
- flour, confectionery and bread normally consumed within 24 hours of preparation;
- vinegar;
- cooking and table salt;
- solid sugar and products consisting solely of flavoured or coloured sugars;

- chewing gums; and
- edible ices in individual portions.

The general requirement of the Food Labelling Regulations 1996 is that the various particulars mentioned, such as the name of the food, list of ingredients, the appropriate durability indication etc should appear:

- on the packaging; or
- on a label attached or the packaging; or
- on a label that is clearly visible through the packaging.

If the food product is sold otherwise than to the ultimate consumer, as an alternative, the details may be on relevant trade documents (except that the name of the food, its appropriate durability indication and the name and address of the manufacturer, packer or seller must appear on the outermost packaging).

Prohibited and restricted claims

Part 3 of the Food Labelling Regulations 1996 deals with claims, nutritional labelling and misleading descriptions in relation to food products as covered by Directive 90/496/EEC. The following claims in the labelling or advertising of a food are prohibited:

(1) A claim that the food has tonic properties (excepting the use of the word "tonic" in the description "Indian Tonic Water").

(2) A claim that a food has the property of preventing, treating or curing a human disease or any reference to such a property.

Schedule 6 to the Regulations outlines both the prohibited claims and restricted claims in relation to food labelling. Restricted claims include:

- claims relating to food for particular nutritional uses;
- reduced or low energy value claims;
- protein claims;
- vitamin claims;
- mineral claims;
- cholesterol claims; and
- nutrition claims.

When considering whether a claim is being made, reference to a substance in an ingredients list or in any nutritional labelling shall not constitute a claim. Taking the example of a protein claim (*i.e.* a claim that a food other than one intended for babies or young children is a source of protein), it must be shown that a reasonable daily consumption of the food would contribute to at least 12 grammes of protein. Foods claimed to be a rich or excellent source of protein must have at least 20% of their energy value provided by protein and in other cases at least 12% and the label for such a food product must give the prescribed nutritional labelling.

Nutritional labelling

Schedule 7 to the Regulations describes the nature of nutritional labelling for foodstuffs. The presentation of the nutritional labelling shall be in one conspicuous place on the label, in a tabular form with the amount in grammes indicated beside the substance. The listing must be in the following format:

Energy
Protein
Carbohydrate, of which
Sugars
Poloyols
Starch
Fat, of which
Saturates
Monounsaturates
Polyunsaturates
Cholesterol
Fibre
Sodium
Vitamins (specific name)
Minerals (specific name)

All amounts should be given per 100 grammes or 100 millilitres and may in addition be given per quantified serving or per portion (if number of portions in pack is stated).

Misleading descriptions

In relation to misleading descriptions they shall not be used in the labelling or advertising of a food, except in accordance with the appropriate conditions set out in Schedule 8 to the Food Labelling Regulations. The words and descriptions covered by Schedule 8 include:

"dietary" or "dietetic", food with an implied flavour, pictorial representation to imply a flavour, "ice-cream", "dairy ice-cream", "milk" or any other word or description with implied milk content; "starch reduced"; "vitamin" or similar description implying vitamins enriched; "alcohol free"; "dealcoholised"; "low alcohol"; "low calorie"; "non-alcoholic"; "liquor" "Indian tonic water" and "tonic water".

To take the example of "low calorie", this description can only be used for soft drink if it contains a maximum of 42 kilo joules (10 kilo calories) per 100 mls. Schedule 8 also includes conditions with regard to the description of cheeses. The following names may not be used in the labelling of any cheese unless the cheese has at least 48% milk fat (dry matter basis) and the cheese has no more than a specified maximum percentage of water: Cheddar, Blue Stilton, Derby, Leicester, Cheshire, Dunlop, Gloucester, Double Gloucester, Caerphilly, Wensleydale, White Stilton and Lancashire. Similar conditions are included in the Schedule with regard to the labelling of creams such as clotted cream, double cream, whipping cream etc.

Penalties

Under regulation 44 if any person who:

(a) sells any food which is not marked or labelled in accordance with the regulations, or

(b) sells or advertises any food in respect of which a claim is made, nutritional labelling is given or description of a name is used in contravention of the Regulations,

that person shall be guilty of an offence and shall be liable on summary conviction to a fine not exceeding £10,000. Under regulation 45 each UK food authority is given powers to enforce and execute these regulations in its own area. In relation to imported food each port health authority shall enforce and execute the Regulations.

A number of sections of the Food Safety Act 1990 shall apply for the purposes of these Regulations, the most important being the sections 21 and 22 defences, in particular, the defence of due diligence. The

Regulations provided for a transitional period, therefore, offences could only be committed under the Regulations as from 1 July 1997.

4.3 Amendments to Food Labelling Regulations 1996

The Food Labelling Regulations have been subsequently amended by a number of statutory instruments. The Food Labelling (Amendment) Regulations 1998 (SI 1998 No 1398) implement Commission Directive 97/4/EEC[2] amending the Framework Directive relating to food labelling. This new Directive is commonly known as "QUID" (quantitative ingredients declaration). In the implementation of QUID, these new Regulations:

- require the quantity of certain ingredients or categories of ingredients of a food to be indicated;
- clarify that where a name of a food is prescribed by EU law, that name must be used as the name of the food and require food brought into Great Britain in certain circumstances from a Member State or an EEA state to comply with details or rules as to the name of the food;
- provide that food consisting of a single ingredient is exempt from the need to carry a list of ingredients only in certain cases;
- require the name "starch" or "modified starch" included in an ingredient list to be accompanied by an indication of its specific vegetable origin and if the starch or modified starch contains gluten;
- provide for these provisions to come into operation on 14 February 2000;
- make a number of other technical and consequential amendments to the 1996 Regulations.

GM food labelling

The Food Labelling (Amendment) Regulations 1999 (SI 1999 No 747) amend the Food Labelling Regulations 1996 to provide for the

---

2　OJ 1997 L43/21.

enforcement of Council Regulation 1139/98[3] concerning the compulsory indication of the labelling of such foodstuffs produced from genetically modified maize and soya. These Regulations require all restaurants, pubs, canteens and catering premises to identify any GM maize or soya ingredients in the foods they sell. The Regulations require all of the United Kingdom's 500,000 catering premises to show which dishes on their menus contain GM soya or maize and ensure that staff, when asked, are able to inform customers which dishes are affected. The products covered are those which are to be delivered as such to the final consumer, having been produced in whole or in part from genetically modified maize or soya.

Under regulation 9 it will be an offence to sell any food to which the labelling requirements of Regulation 1139/98 apply without marking or labelling the actual presence of GM maize or soya. Persons who do not comply with the various provisions of this regulation may be fined up to £5,000 for non-compliance. The Regulations provide for a number of sections of the Food Safety Act to apply in relation to these Regulations and in particular the section 21 defence of due diligence. These Regulations came into force on 19 September 1999.

The Food Labelling (Amendment) (No 2) Regulations 1999 (SI 1999 No 1483) amend the Food Labelling Regulations 1999 in order to implement Commission Directive 1999/10/EEC[4] providing for derogations from the provisions of Article 7 of Council Directive 79/112/EEC regarding the labelling of foodstuffs. The Food Labelling Regulations 1996 require the quantity of certain ingredients or categories of ingredients of a food to be indicated. These new Regulations:

- remove that obligation in the case of sweeteners, sugars, vitamins or minerals used in the preparation of a food in certain circumstances;
- provide some derogation's from the existing method for calculating the quantity of ingredients or category of ingredients; and
- make various other consequential amendments and ensure that these provisions will come into force from 14 February 2000.

These Regulations also require prepacked food sold or supplied as an individual portion and intended as a minor accompaniment to another

---

3   OJ 1998 L159/4.
4   OJ 1999 L69/22.

food to be marked or labelled with details relating to packaging gases, added sweeteners or added sugars unless this has been exempted under the Food Labelling Regulations 1996.

Finally, the Food Additives Labelling Regulations 1992 (SI 1992 No 1978) which came into force on 14 September 1992, implement Articles 7 and 8 of Council Directive 89/107/EEC on the approximation of the laws of Member States concerning food additives authorised for use in foodstuffs intended for human consumption. The principal provisions of these Regulations:

- define food additives; and
- prescribe labelling requirements for sales of food additives.

The provisions of these Regulations do not apply to any food additive once it has become part of another food. If any person contravenes or fails to comply with any of the provisions of these Regulations they shall be guilty of an offence and liable on summary conviction to a fine not exceeding £5,000.

4.4 **Case law**

Misleading claims

The making of misleading claims about the medicinal qualities of food products is prohibited in both the Food Safety Act 1990 and the Food Labelling Regulations 1996. The issue was dealt with in *Cheshire CC* v *Mornflake Oats Ltd.*[5] The company placed an advertisement in *The Times* which contained a claim stating:

> "Research has proved that oats can help to reduce excess cholesterol levels when eaten as part of a low-fat diet, thereby cutting down the risk of heart disease ... Find out how one of nature's most enjoyable and versatile foods can be a key to preventing heart disease."

The company was charged with making a claim in contravention of the Food Labelling Regulations 1994 (superseded by the 1996 Regulations). The prosecution contended that if the advertisement was taken as a whole, there was an express or implied claim that Mornflake Oat's products were capable of preventing heart disease. The company argued that the advertisement was part of a special book promotion

---

5   *Solicitors Journal*, 19 Nov 1993.

and therefore the implied health claim did not exist. The court decided that it was indeed an ambiguous advertisement and did include an implied health claim, therefore, Mornflake Oats Ltd were asked to discontinue this form of advertisement.

Another misleading claim case, *Director of Fair Trading v Tobyward Ltd*[6] concerned the slimming aid "Speedslim", while *R v Warwickshire County Council ex parte Johnson*[7] related to misleading price promotion claims which had important implications for the food industry as it ruled that price promotions must be honoured and it was no defence to plead that all reasonable efforts were taken to supply the food product part of the promotion campaign.

London Borough of Hackney v Cedar Trading Ltd[8] is a recent case involving the Food Labelling Regulations 1996. Cedar Trading purchased consignments of Sprite and Coca-Cola made in Holland, where it had been lawfully sold and distributed them through Pelin Supermarket in the United Kingdom. The drink cans pictorially looked the same as similar products made in the United Kingdom, but information on the labelling as to the nature of the drink, *i.e.* "soft drink with vegetable extracts" was in Dutch as was the list of ingredients. The Food Labelling Regulations 1996 state that the name used for a foodstuff and the list of ingredients must be in a "format" understandable to the purchaser. The magistrates' court asked the divisional court whether a well-known brand name could confer an exception to these obligations on the name/ingredients contained in the Food Labelling Regulations 1996. The divisional court ruled that a well-known brand name did not constitute an "exception" to the Regulations. Normally, such a finding would be a direction to the magistrates' court to continue with the case, but as Cedar Trading was in liquidation the court said the judgment would mark the end of the matter. This judgment is in line with a recent European Court case, *Criminal Proceedings involving Goerres*[9] which ruled that all compulsory particulars (*e.g.* name, list of ingredients etc) specified in Directive 79/112/EEC on food labelling must appear in a language easily understood by consumers of the state or region in question.

---

6   [1989] 2 All ER 266.
7   House of Lords, 10 Dec 1992.
8   *Food Law Monthly*, Aug 1999.
9   Case C-385/96.

4.5 **Food labelling legislation in Ireland**

In Ireland the Framework Directive on Food Labelling 79/112/EEC has been implemented by the European Communities (Labelling, Presentation and Advertising of Foodstuffs) Regulations 1982 (SI 1982 No 205). These Regulations require pre-packaged foods, subject to various exceptions specified in the Regulations, to be marked or labelled with:

- the name of the foodstuff;
- a list of ingredients;
- the net quantity;
- the date of minimum durability;
- any special storage conditions;
- the name and address of the manufacturer, packer or seller;
- particulars of the place of origin and instructions for use, where the absence of this information would mislead the consumer.

These labelling items therefore align themselves with conditions laid down in Article 3 of the EU Directive. Regulation 6 of the 1982 Irish Regulations specifies that the labelling provisions do not apply to the sale of pre-packaged foodstuffs that are packaged by a person selling the foodstuffs on the retail premises, *e.g.* in a confectionery shop or a vegetable shop. There are conditions laid out in the Regulations that in such situations a number of specific items requested under the 1979 Regulations should be displayed in a notice in the premises in order that consumers would not be misled about the foodstuff products they are purchasing.

Exemptions

The Food Labelling Regulations 1982 also shall not apply in relation to:

- foodstuffs whose minimum durability exceed 18 months (*e.g.* honey);
- deep frozen foodstuffs;
- ice-creams;
- chewing gums and similar chewing products;
- fermented cheese intended to ripen completely or partially in pre-packaging.

Similarly, the Irish Food Labelling Regulations do not apply to chocolate confectionery products and sugar confectionery products which have a net weight of less than 50 grams.

Under regulation 14 an authorised officer may at all reasonable times enter any premises in order to ascertain whether particular food products are being sold in line with the 1982 Regulations. During such an inspection the authorised officer may take copies or extracts from any books, records and documents available in the food premises and also take such samples of any foodstuffs or materials, which he may deem fit. The person in charge of the food premises must afford the authorised officer all facilities and assistance which are reasonably necessary in order to enable the officer to perform his functions under the terms of the 1982 Regulations. Where a sample is taken, the authorised officer concerned must divide the sample into not more than four approximately equal parts, each of which he shall mark in such a way as to identify it as part of the sample taken by the officer. In any proceedings taken for an offence under these Regulations one of these samples must be transmitted to the defendant before any court proceedings.

Penalties

A person who obstructs or interferes with an authorised officer in the course of exercising powers conferred on him under the 1982 Regulations shall be guilty of an offence. Under the 1982 Regulations a person who sells foodstuffs in contravention of the Regulations shall be liable upon summary conviction to a fine not exceeding £800 or at the discretion of the court to imprisonment for a term not exceeding six months or to both a fine/imprisonment as the court deems fit. A person guilty of an offence under regulation 14 (obstructing an authorised officer) shall be liable upon summary conviction to a fine not exceeding £500.

Defences

Regulation 16 provides defences for those prosecuted under these provisions where they can show that they received the foodstuffs in the knowledge that the products were in compliance with the 1979 Food Labelling Directive and the Irish Labelling Regulations 1982. It will not be a defence for them to show that they had a written warranty to the effect that the foodstuffs were labelled correctly and that the person prosecuted took all reasonable precautions and exercised all due diligence to avoid the commission of any offence by him or any other person under his control. The Regulations specify that a statement by the manufacturer, importer or seller of foodstuffs in an invoice or a label attached to the foodstuffs stating that the foodstuffs comply with the provisions of the 1979 Food Labelling Directive and the 1982 Food Labelling Regulations shall be deemed for the purposes of a food retailer to be a warranty against prosecution. A person shall not without leave of the court be entitled to rely on these defences unless seven days before the court hearing he has served on the prosecutor notice in writing giving his intentions that he proposes to rely on that defence. An offence under the 1982 Regulations may be prosecuted by the minister or by a health board in whose functional area the offence was committed. A person guilty of an offence under these Food Labelling Regulations is liable to a fine on summary conviction not exceeding £1,500.

Amendments

The 1982 Regulations have been subsequently amended on a number of occasions. The European Communities (Labelling, Presentation and Advertising of Foodstuffs) (Amendment) Regulations 1995 (SI 1995 No 379) give effect to the provisions of Commission Directives 93/102/EEC and 94/54/EEC. Under the 1995 Regulations categories of ingredients will now have to be designated according to the revised Annexes of Directive 93/102/EEC. Annex 1 lists the categories of ingredients which may be designated by the name of the category rather than the specific name, for example: starches and starches modified by physical or enzymatic means; all fish species; different types of poultry meat are all defined and given a simple designation – starch, fish and poultry meat – which can be used on the food label. Annex 2 lists the categories of ingredients which must be designated by the name of category to which they belong followed by their specific name or EEC number, for example, colour E211, stabiliser E453, preservative E176 etc. In relation to the labelling of foodstuffs whose durability has been extended by means of packaging gas the 1995 Regulations also specify that the label must contain the indication "packaged in a protective atmosphere"

The most recent amendment to the 1982 Regulations is the European Communities (Labelling, Presentation and Advertising of

Foodstuffs) (Amendment) Regulations 1997 (SI 1997 No 151). The effect of these new Regulations will be to require specific indication, on the labelling of foodstuffs, as to the presence of sweeteners and/or added sugars. The presence of particular sweeteners such as aspartame and polyols which may cause adverse effects on some people must now be indicated on the food label.

Finally, in relation to the nutritional labelling of foodstuffs, Council Directive 90/496/EEC on nutritional labelling of foodstuffs has been implemented into Irish legislation by the Health (Nutritional Labelling of Foodstuffs) Regulations 1993 (SI 1993 No 388). These Regulations define in detail nutritional labelling and the nutritional claims that can be used on a food label. The Regulations do not apply to natural mineral waters, diet integrators or food supplements. No nutritional claim can appear on a food label other than those relating to energy, to nutrients (protein, carbohydrates, fat, fibre, sodium and various vitamins and minerals listed in the Schedule to the Regulations) or to substances which belong to or are components of a category of the nutrients previously listed.

Enforcement

An authorised officer may at all reasonable times enter any premises to ensure that the 1993 Regulations are being properly implemented. Any person found guilty of contravening the Nutrition Labelling Regulations may be prosecuted by a health board within the functional area of which the offence was committed and will be prosecuted under the provisions of the Health (Official Control of Food) Regulations 1998.

4.6 Conclusions

Food labelling is one of the main areas of food law and the food label plays an important role in communicating vital information to the consumer about different food products. Unfortunately for food manufacturers and retailers, labelling is never static as there are continuous demands for more and different labelling information. There are demands in the United Kingdom for food labelling to cover health claims, GMOs, food allergies, environmental claims, *e.g.* "dolphin-free tuna", organic labelling etc. The simple fact though is that there is only so much space on any specific label, and even in future where symbols are used, there is a case to be made that we have reached saturation-level in terms of what can be placed on the food label. Legislators and consumer groups too often take the easy option by advocating "put the information on the label". With the advent of the "Information Society", it must be possible to provide information to consumers via toll-free numbers, internet sites and brochures in supermarkets.

The UK Co-op produced a report in November 1997 "The Lie of the Label" which showed that consumers were being duped by inaccurate labels which too often failed to reveal the whole truth about the content of food products. As there is sufficient UK legislation on food labelling, it is now essential to make sure that it is properly enforced. Food labelling will continue to be one of the main tools whereby legislators ensure that consumers have confidence in the food products they purchase. The hope must be that the UK government will make the effort to enforce the law in this area, even if, unlike food hygiene issues, it does not seem to have the "immediacy" factor in relation to the question of overall food safety.

Chapter 5

· **Genetically Modified Foods** ·

Summary

- UK law regarding genetically modified foods
- Novel Foods Regulations 1997
- Liability issues in relation to GM foods – case law
- Genetically Modified Food and Producer Liability Bill 1999
- Proposes introducing a strict liability regime for damaged caused by genetically modified organisms (GMOs)
- GM food law in Ireland
- US class action cases in relation to GM foods filed in December 1999

5.1 **Introduction**

The current controversy over genetically modified foods has caused food manufacturers/retailers to fall over each other trying to prove to consumers that their products are safe and free of GMOs. The biotechnology industry has few problems persuading consumers about the advantages of gene technology with regard to new medical remedies for various diseases, but when it comes to food, consumers are generally sceptical about the advantages of this technology to the production of foodstuffs. The BSE crisis has undermined consumers' confidence in relation to scientific pronouncements of "safety" in relation to food issues. GM foods have little future in Europe until this faith can be restored. The EU has a bad record in this area of food law in its attempts to try and introduce a uniform labelling system for genetically modified foods. As to questions of liability in relation to the introduction of gene technology and genetically modified foods into the food chain, the EU has shown no leadership and therefore it looks likely national governments will have to tackle these issues themselves, without any guidance from the European Union.

5.2 **UK legislation on GM foods**

The European dimension

There are two main pieces of European legislation, covering the whole issue of genetically modified foods.

1. Directive 90/220/EEC on GMOs

Council Directive 90/220/EEC[1] is on the deliberate release into the environment of genetically modified organisms (GMOs). **Genetically modified organisms** are defined as organisms in which the genetic material has been altered in a way that does not occur naturally by mating and/or recombination. In these cases, new materials are "recombined" into the DNA structure of living cells, for example, perishable material prepared outside the organism. These organisms can consist of seed from maize, corn, etc which at a later stage would be used ultimately in the production of foodstuffs. In the United Kingdom, this Directive has been implemented by the Environmental Protection Act 1990. Under this Act an authorisation procedure has been established whereby an undertaking or a company is required to submit a notification to the Department of the Environment whether initially to have a GM trial or actually to obtain authorisation for a GMO product to be introduced into the environment.

The 1990 Directive is presently undergoing amendment in the European Union and on 25 June 1999 the Environment Council reached agreement by qualified majority voting on a proposal to amend Directive 90/220/EEC concerning the release of GMOs into the environment. The common position adopted strengthens the Commission's proposal by imposing further safety guarantees, improving the transparency of the authorisation procedure and addressing consumer concerns by introducing compulsory labelling and public consultation. The amendments will now go to the European Parliament for a second reading. However, two groups of Member States made political declarations that they would not grant any further authorisations until the amended Directive comes into force in the year 2002. These groups are of sufficient size to block further authorisations with a result that a *de facto* moratorium on the release of GMOs will continue in the European Union for the foreseeable future.

---

1   OJ 1990 L117/15.

2. Regulation 258/97 on novel foods

EP and Council Regulation 258/97[2] on novel foods and novel food ingredients provides for an authorisation procedure in order to place a novel food product on the market. Foods falling within the scope of this Regulation according to Article 3 must not:

- present a danger for the consumer;
- mislead the consumer; and
- differ to such an extent from "traditional" foods that they would be nutritionally disadvantageous for the consumer.

In order to place a novel food product on the market a request and technical dossier must be provided both to the competent national authority as well as to the European Commission at the same time. After a long process of consultation between the Member States and the European Commission a final decision is made regarding the approval of the novel food. Two statutory instruments introduced by means of the Novel Foods Ingredients Regulations 1997 (SI 1997 Nos 1335 and 1336), provide for the enforcement and execution of the various provisions of the EU Novel Foods Regulation in the United Kingdom. These Regulations which came into force on 16 June 1997 designate the Minister of Agriculture Fisheries and Food and the Secretary of Health to act jointly as the food assessment body for the purposes of Regulation 258/97. Certain provisions of Regulation 258/97 are specified in a Schedule to the statutory instruments.

The UK Regulations require that novel food and food ingredients falling within the scope of the EU Regulation must not be dangerous, misleading or differ from the foods or food ingredients that they are intended to replace so as to be nutritionally disadvantageous. Those wanting to place such a food product on the market are requested to submit a request to the Member State in which the product is to be first marketed and at the same time to submit a copy of the request to the European Commission. Labelling requirements ensure that the final consumer must be informed of any characteristic or food property, which renders a novel food or novel food ingredient no longer equivalent to an existing food or food ingredient. In total, there are eight different requirements listed in the Schedules to these Regulations and any person who contravenes or fails to comply with any of these requirements/specified provisions shall be

---

2   OJ 1997 L43/1.

guilty of an offence and can be held liable on summary conviction in court to a fine not exceeding £5,000.

In the second of the statutory instruments dealing with novel foods and novel foods ingredients, a scale of fees is established for payment to the Minister of Agriculture Fisheries and Food when a food manufacturer submits a request to him or her relating to a novel food or novel food ingredients. These fees are as follows:

(1) In respect of a request relating to an application involving genetically modified organisms for which an environmental risk assessment is required - £6,500.

(2) In respect of a request relating to an application to place a novel food or novel food ingredient on the market - £4,000.

(3) In respect of a request for an opinion as to the substantial equivalence of a novel food with an existing food product - £1,725.

Finally, the Regulations stipulate that each local authority in the United Kingdom is given the authority to enforce and execute the provisions of Regulation 258/97.

5.3 Liability issues

In recent times there has been considerable debate over liability in relation to GMOs and genetically modified foods. It is interesting to note that on 3 February 1999, the UK Minister of State in the Ministry of Agriculture Fisheries and Food, Geoff Rooker, in reply to a question concerning liability in the area of GM foods stated:

> "Civil liability for damage caused by genetically modified organisms is covered by the common law developed in the courts. On the basis of common law principles, the firm holding the marketing consent for the GMO crop can be held liable in law for any damages arising from ill effects attributed to that crop."[3]

Producers and importers of most products have long been held liable for any damage caused to consumers by defects in their products under the EU Directive on Product Liability. As stated previously, the Product Liability Directive originally applied only to processed foods but has now been extended to cover primary agricultural products and game.

---

3   Hansard Official Report, 3 Feb 1999, Vol 324, c 864.

The issue of liability was considered by the EU Environment Ministers in June 1999 when debating a revision of Directive 90/220 on the deliberate release of GMOs into the environment. The UK government issued a statement at that time calling on the European Commission as a matter of priority to consider, outside the framework of Directive 90/220/EEC, the feasibility of the introduction of a liability regime to cover the release and marketing of GMOs. The law on liability for the release of GMOs is therefore far from clear.

If cross-pollination occurs and a new hyber plant is created, *e.g.* a weed that is resistant to conventional herbicides, the extent of environmental damage is completely unpredictable, but could be enormous. One result is that insurance companies are facing serious problems in designing policies to cover companies involved in GM food and crops. Despite the UK government's request to consider developing an EU-wide liability regime to cover the release and marketing of GMOs, at present in the United Kingdom liability for environmental damage is covered by the normal rules of tort and there have not yet been any cases to determine how these rules should apply to GMOs. An example of a potential liability issue could be where pollen from a GM crop grown in a farmscale trial contaminates a nearby organic farm, threatening that farm's organic certification. Whom should the organic farmer sue? The GM company that designed the crop or the farmer who grew it, even if that farmer has followed all environmental advice to minimise the risk of cross-pollination? In the event of any health damage from GM food, liability might be faced by the GM company, the food manufacturer who has used GM ingredients, or the retailer who sold the food to the consumer?

Case law

The only case concerning this issue is not very clear on the liability issues involved. In *R v Secretary of State for the Environment, Transport and the Regions and another ex parte Watsons and Sharpers International Seeds Ltd*[4] an organic farmer fearing that there might cross-pollination of his crop by a nearby GM field trial, sought to have the GM crops destroyed before they caused contamination to his crop. The Court dismissed his action as the judges believed that all necessary precautions had been taken and that they could not deal with threats to

---

4   *The Times* Law Report 31 Aug 1998, CA.

the organic status of the applicant's produce before the event took place, especially as it was their view that the GM field trial had followed the existing UK licensing procedures.

5.4 **Genetically Modified Food and Producer Liability Bill**

This Bill was presented to the House of Commons on 24 June 1999, as a Private Member's Bill. This Bill amends section 111(8) of the Environmental Protection Act 1990 in order to introduce the precautionary principle as a means of risk assessment for genetically modified organisms. The "precautionary principle" means that where there is a risk of significant damage to human health or the environment, lack of scientific certainty should not be used as a reason for not taking or for postponing measures to avoid or minimise such a risk.

Clause 2 of the Bill introduces a strict liability regime for damage caused by genetically modified organisms. Under this clause a person who holds a consent, as provided for under section 111 of the Environmental Protection Act 1990, or who holds consent given by another Member State under Article 13(4) of Council Directive 90/220/EEC, would be liable for any damage which is caused by the deliberate release or marketing of a genetically modified organism under the terms of that consent.

Damage is defined under clause 3 of the Bill as:

(a) Personal injury;

(b) Damage to property;

(c) Financial loss;

(d) The cost of protecting against preventing, remedying or rectifying environmental damage; and

(e) Damage to the environment within the meaning of Article 2.10 of the Lugano Convention on Civil Liability for damage resulting from activities dangerous to the environment.

Clause 22 ensures that liability may be incurred by a body corporate unless it can show that it did everything in its power to prevent the deliberate release or marketing which caused the damage in question. Where damage to the environment occurs outside the meaning of various situations defined in (a) to (d) above, the Secretary of State or with the leave of the court, any other person may apply to the court for damages to be awarded against a potential defendant. In reaching its decision on such an application the court may have regard to such matters as:

(1) The severity and detrimental effect of the damage on the environment.

(2) Any relevant profits made by a potential defendant.

(3) Any relevant remuneration received by a potential defendant.

In any proceedings taking place under clause 3 of this Bill it would be for the person proceeded against to prove that he did not cause the damage in question and where proceedings are brought against more than one person, it would not be a requirement for the person bringing the proceedings to identify the person who caused the damage in question, provided that he can prove that one or more of the persons so proceeded against could have caused the damage.

There is a defence available under clause 4(3) if the person proceeded against can prove that the damage in question was caused by an exceptional case of *force majeure*. Clause 5 of the Bill concerns the various indemnities that a potential defendant should indemnify against, while clause 6 introduces the statutory requirement that all potential defendants should be required to take out a policy of insurance against liability.

Penalties

A person found guilty of an offence under the Act or of failing to take out appropriate liability insurance, on summary conviction would be liable to a fine not exceeding £1,000 and on conviction on indictment to a fine not exceeding £5,000 or a term of imprisonment not exceeding three months or both. Finally, clause 7 provides for the establishment of a genetically modified organism "compensation fund" which would provide for the payment of compensation in respect of damage caused by deliberate releases or marketing of GMOs where liability for damage cannot be attributed to an identifiable potential defendant.

5.5 GM foods legislation in Ireland

The Minister for the Environment has overall responsibility for the implementation of Directive 90/220/EEC on the deliberate release of

GMOs into the environment. The Directive is implemented into national law by the Genetically Modified Organisms Regulations 1994 (SI 1994 No 345) and the Environmental Protection Agency (EPA) is the competent authority for administering the legislation at national level. It is also the designated authority to administer and enforce Novel Foods Regulation 258/97. All licence and consent applications must be made to the EPA.

In October 1999, the government approved a precautionary environmental policy on GMOs based on scientific risk assessment and management, following the publication of a consultation paper on the subject. The main recommendations of the consultation paper were:

- a review of the Environmental Protection Agency's resources;
- the inclusion of consumer representation on the Agency's GMO Advisory Committee;
- a research programme on environmental safety to be identified and managed by the EPA; and
- a greater effort by all state agencies concerned to provide adequate information to the public.

Finally, the Irish government has brought forward no proposals regarding the establishment of a special liability regime covering GMOs and genetically modified foods.

5.6 Conclusions

The question of liability in relation to genetically modified foods is likely to become a major news item in 2000. The US law firm, Cohen, Milstein, Hausfeld & Tolls announced in December 1999, that it is launching a series of class actions similar to the present tobacco cases, in which it will demand millions of dollars of damages from the principal companies involved in the production of genetically modified seeds and crops. The targets of these class actions brought by farmers in the United States, the European Union, Central America and India, is likely to include large biotech companies such as Monsanto, Du Pont, Astra Zeneca, Novartis and Agr-Evo. The class actions will allege "anti-competitive behaviour" in the seed market, which is dominated by a small number of companies, in violation of anti-trust laws. It will cite "questionable corporate behaviour" in pushing forward with the rapid introduction of GM foods into the food production chain in absence of clear data to prove their safety. The class actions not only seek damages, but the hope is that the legal proceedings will put the deployment of GM foods on hold until their safety is scientifically proven. It also hopes to establish legal liability where farmers were burdened with unmarketable crops that were either grown from GM seed or contaminated with GM material from neighbouring fields.

As this chapter indicates, there are moves in the United Kingdom to introduce a liability regime to cover GM seeds and GM foods. The Private Member's Bill, The Genetically Modified Food and Producer Liability Bill is one such example, but at EU Environment Council Meetings the UK government has also been advocating that the EU devise a liability regime for GM seeds and plants. The US class actions will concentrate the minds of UK and European legislators to look at this issue with extreme urgency. The EU's record so far in relation to the labelling of GM foods is not good and it has seemed at times that science is moving faster than the legislators. The new EU Commissioner for Health and Consumer Protection, David Byrne, has promised to give this issue his immediate attention. If he neglects to tackle this problem, he will not only face problems with irate consumers but may also potentially obstruct the functioning of the internal market as well as face a possible trade war with the United States.

Chapter 6

Regulations for Different Food Products

Summary

- Compositional laws in relation to specific foodstuffs
- Cocoa and Chocolate Products Regulations 1976
- Natural Mineral Waters, Spring Water and Bottled Drinking Water Regulations 1999
- Foodstuffs for particular nutritional uses – processed cereal foods and infant & follow-on formulae
- Compositional food law in Ireland.

6.1 Introduction

Areas such as food labelling, food hygiene and food additives lend themselves to specific legislation covering numerous food products, which may ultimately be sold to the consumer. On the other hand, there is an existing body of legislation covering specific food products both in terms of their composition and quality standards. The main area of food law where the concept of quality was introduced in the past has been in the context of compositional standards for various foodstuffs. Many EU Member States continue to have compositional standards legislation on their statute books for a whole range of foodstuffs. Since 1985, in the wake of the *Cassis de Dijon* judgment, the European Union no longer establishes legislation in terms of compositional standards, rather it has given priority to the establishment of horizontal legislation covering issues such as consumer protection and fair trading, *e.g.* the Framework Labelling Directive and the Framework General Hygiene Directive.

Despite this, there are areas where the European Union has in more recent times introduced what is termed as vertical legislation, which has a compositional basis. This chapter will cover a number of pieces of legislation, which existed prior to 1985 and have been implemented into

UK law – the examples taken will be chocolate and natural mineral waters. On the other hand, in relation to foods for particular nutritional uses, the European Union established a Framework Directive on such foodstuffs in 1989 and since then it has introduced specific legislation which has a compositional make-up on subjects such as infant and follow-on formulae, processed cereal based foods and foods for weight reduction. This chapter will look at the implementation of the processed baby foods and infant and follow-on formulae Directives into UK law. These examples will show that there is a body of EU/UK food law, which is much more descriptive in terms of the obligations it imposes on the food manufacturer/retailer. Obviously, there are numerous other food products which are covered in this way but questions of space mean that a few examples will be covered in this chapter which will give the reader an idea of how this type of law is drafted.

6.2 Cocoa and chocolate products

Commission Directive 73/241/EEC[1] on cocoa and chocolate products has been implemented into UK law by the Cocoa and Chocolate Products Regulations 1976 (SI 1976 No 541). The EU Directive defined a number of cocoa and chocolate products, such as cocoa powder, drinking chocolate, chocolate, plain chocolate, chocolate flakes and milk chocolate. Each definition consisted of the compositional/quality requirements necessary in order that that product could be correctly labelled, *e.g.* "drinking chocolate" must consist of a mixture of cocoa powder and sucrose so that every 100 grams of the mixture contains 32 grams of cocoa powder. Particular labelling requirements were also included besides the composition of the product.

This Directive has caused much controversy in the United Kingdom and Ireland due to its definition of cocoa and chocolate products, by making no provision for the use of vegetable fats to replace cocoa butter in the production of chocolate. Following the accession of the United Kingdom and Ireland and on the basis of rules applicable in those countries, an exemption was introduced permitting the use of vegetable fats up to a limit of 5% in the manufacture of chocolate. Subsequent attempts by the European Commission to authorise this 5% limit derogation have run into great difficulty and a proposal on this matter is still before the Internal Market Council. According to the

---

1   OJ 1973 L228/23.

text the use of non-cocoa butter vegetable fats will be permitted up to a limit of 5%, above that limit the chocolate product would have to include a statement within the ingredients list stating "contains vegetable fats in addition to cocoa butter". Chocolate products with more than 20% milk content, which are produced mainly in the United Kingdom and Ireland will also have to be labelled "family milk chocolate" for the export market, although that would now apply to domestic sales. This proposal has still to be adopted at EU level.

Definitions

The Cocoa and Chocolate Products Regulations 1976, give various definitions for chocolate products, chocolate, milk chocolate etc. Just as the EU Directive defined the various compositional standards for such products, the UK Regulations contain a long Schedule defining both cocoa and chocolate products and their compositional make-up. Schedule 1, Part 1, therefore, deals with cocoa products, *e.g.* "drinking chocolate" is defined as being "a mixture of cocoa and sucrose containing less than 20% cocoa". On the other hand, Part 2 of Schedule 1 refers to chocolate products, *e.g.* chocolate is specified as being:

> "any product obtained from cocoa nib, cocoa mass, cocoa, fat reduced cocoa or any combination of two or more thereof and sucrose with or without the addition of extracted cocoa butter and containing not less than 35% of total dry cocoa solids, including not less than 14% dry non-cocoa solids and not less than 18% permitted cocoa butter."

The Regulations specify that all cocoa and chocolate products may contain:

- any miscellaneous additive (subject to the Miscellaneous Food Additives Regulations 1995[2]); and
- any flavouring substance which does not impart the flavour of chocolate or milk fat.

Labelling requirements

The labelling and description of cocoa and chocolate products is covered by regulation 5 of the Regulations and it specifies, in the case of

2   SI 1995 No 3187.

chocolate, that the percentage of total dry cocoa solids, non-fat cocoa solids and cocoa butter must be referred to on the label of a chocolate product, as well as the name or trade name and the address or registered office of the manufacturer or packer of the cocoa or chocolate product or a seller established within the European Union.

Penalties and enforcement

If any person contravenes or fails to comply with the various compositional standards/labelling requirements contained within the Regulations, he shall be guilty of an offence and shall be liable to a fine not exceeding £100 or to imprisonment for a term not exceeding three months or both and in the case of a continuing offence to a further fine not exceeding £5 for each day during which the offence continues after conviction. Local food authorities are nominated as the enforcement bodies under these Regulations.

Amendments

The Cocoa and Chocolate Product Regulations were amended in 1982 by the Cocoa and Chocolate Products (Amendment) Regulations 1982 (SI 1982 No 17). These new Regulations amend the definition of "milk solids", "gianduja nut chocolate" and make minor alterations to the amounts of various flavourings and emulsifiers which may be used to produce these chocolate products.

6.3 Natural mineral waters

The European Union introduced Commission Directive 80/777/EEC[3] to establish common rules on the microbiological requirements and the conditions whereby specific names could be used for certain types of natural mineral waters. Such waters would be recognised by a responsible national authority, which had the task of instigating regular checks to ensure that the natural mineral water producer was following the various provisions of the Directive. Each Member State must notify the Commission of the various natural mineral waters it has recognised under the Directive, after which, in order to improve

3   OJ 1980 L229/1.

transparency they will be published in the *Official Journal of the European Communities.*

This Directive was implemented in the United Kingdom by the Natural Mineral Waters Regulations 1985 (SI 1985 No 71) which have recently been revoked and superseded by the Natural Mineral Waters, Spring Water and Bottled Drinking Water Regulations 1999 (SI 1999 No 1540). Under these new Regulations a "**natural mineral water**"means water which:

> "(a) Is microbiologically wholesome;
>
> (b) Originates in an underground water table or deposit and emerges from a spring tapped at one or more natural or bore exits;
>
> (c) Has not been subject to any treatment or addition other than the separation of its unstable elements such as iron and sulphur compounds and the total or partial elimination of free carbon dioxide; and
>
> (d) Is recognised as being a natural mineral water following an application to the relevant national authority."

Under regulation 4 there is provision for the relevant authority (local council/food authority) to decide initially whether to grant recognition of the water for the purposes of these Regulations. Subsequently, if after reviewing the decision and instigating various investigations the relevant authority has the power to withdraw the recognition of the natural mineral water if, *e.g.* the water is not in line with the prescribed concentrations or values of parameters required for natural mineral waters in the various Schedules to the 1999 Regulations. It is prohibited under regulation 5 for any person to sell water, the marking or labelling of which uses the name "natural mineral water" unless the water has been recognised as such by the competent national authority.

The Schedules to the Regulations introduce a normal viable colony count, which is necessary for any natural mineral water to possess. Where it is found during exploitation that the natural mineral water is polluted and the bottling or sale of the water would be in contravention of these Regulations, the spring from which the water is extracted must not be exploited nor shall the water be bottled until the cause of pollution is eradicated. Under regulation 9 natural mineral water should be bottled in a container which is fitted with closures designed to avoid any possibility of adulteration or contamination.

Labelling requirements

Besides the term "natural mineral water" being displayed on the label, the following information must also be included:

"(a) A statement of the analytical composition which shall indicate the characteristic constituents of the water;

(b) The name of the place where the spring is exploited and the name of the spring;

(c) Where the water has undergone the total or partial elimination of free carbon dioxide by exclusively physical methods, the indication "fully decarbonated" or "partially decarbonated" should be used."

Spring water and its compositional make-up is described in regulation 11 and the same procedures are defined in terms of bottled drinking water in regulation 12.

Enforcement

Under the Regulations each food authority shall enforce and execute these Regulations within its own area. The relevant food authority can carry out periodic checks on any water which has been recognised as a "natural mineral water" in order to ensure that it is following all the obligations necessary under the 1999 Regulations.

Samples procedure

An authorised officer of a food authority can procure a sample of a natural mineral water using powers contained in section 29 of the Food Safety Act 1990, which apply in the context of these Regulations. The sample that the authorised officer must procure should include at least one or more bottles of any natural mineral water. The sample shall be divided into three parts, each part to be marked and sealed or fastened, one part of the sample which will be returned to the person from whom it was purchased or in the case of water being brought into the United Kingdom to the person who intends to sell that water in the United Kingdom, or if neither of these cases to the person appearing to be the owner of the water from which the sample was taken. The person receiving the sample shall be given notice that the other two samples will be sent for analysis by a public analyst under the conditions laid down in section 30 of the Food Safety Act 1990. These samples can then be provided to the government chemist or to a court as specified so as to bring a case against a person guilty of selling water which is not a natural mineral water or spring water or bottled drinking water under the terms of these Regulations.

Penalties

A person guilty of such an offence or a person who has refused to provide samples to an authorised officer can be liable on summary conviction to a fine of £5,000. A defence is available for the person charged if he can prove that the water in respect of which the offence was alleged to have been committed was intended for export to a country which has legislation analogous to these Regulations and it complies with that legislation.

6.4 Foodstuffs for particular nutritional uses

Under EU law there is a specific framework Directive[4] approximating the laws of the Member States relating to foodstuffs intended for particular nutritional uses. The Directive establishes a common definition for such foodstuffs which:

> "Owing to their special composition or manufacturing process are clearly distinguishable from foodstuffs for normal consumption, which are suitable for their claimed nutritional purposes and which are marketed in such a way as to indicate such suitability."

The particular nutritional use must fulfil one of the following requirements:

(1) It must be prepared for certain persons whose digestive processes or metabolism is disturbed.

(2) It must be prepared for persons in a special physiological condition and who derive special benefit from consuming such foodstuffs.

(3) It must be prepared for infants or young children in good health.

A number of specific Directives have been introduced under the framework of this Directive relating to baby foods and these will now be discussed in some detail.

---

4   Council Directive 89/398, OJ 1989 L186/27.

6.5 **Processed cereal based foods**

European Commission Directive 96/5/EEC[5] on processed cereal-based foods and baby foods for infants and young children has been implemented by the Processed Cereal Based Foods and Baby Foods for Infants and Young Children Regulations 1997 (SI 1997 No 2042). These Regulations cover foodstuffs for use by infants while they are being weaned and by young children as a supplement to their diet and/or for their progressive adaptation to ordinary food.

Processed cereal-based foods include simple cereals which can be reconstituted with milk, cereals with an added high protein content which are reconstituted with water and rusks/biscuits used either directly or with the addition of milk/water. Processed cereal-based foods and baby foods must be manufactured from ingredients whose suitability has been established by generally accepted scientific data.

Composition

The compositional criteria for the manufacture of processed cereal-based foods and baby foods are contained in Part 2 of Schedule 1 to the 1997 Regulations. In terms of cereal content, *e.g.* processed cereal-based foods must be prepared primarily from one or more milled cereals and/or starchy root products. The amount of cereal and/or starchy root must not be less than 25% of the final mixture on a dry weight for weight basis. Other compositional criteria are included for proteins, carbohydrates, fats, and minerals including sodium and calcium and for various vitamins. The compositional criteria for baby foods are set out in Schedule 2 to the Regulations and Schedule 4 lists any added nutritional substances which may be used in the manufacture of either of these types of foodstuffs.

Labelling requirements

In relation to the labelling of processed cereal based foods and baby foods the following particulars must be taken into account:

(a) A statement must be provided as to the appropriate age (which shall not be less than four months) from which the food may be used.

---

5   OJ 1996 L49/17.

(b) Information should be provided on the label as to the presence or absence of gluten if the product is for a baby aged less than six months.

(c) The label should include the available energy value expressed in kJ and kCal and the protein, carbohydrate and fat content, expressed in numerical form per 100 grams or 100 mls of the food as sold and where appropriate per specified quantity.

(d) The average quantity expressed in numerical form per 100 grams or 100 mls of the food as sold of each mineral substance and of each vitamin in respect of which a maximum or minimum compositional requirement is specified in the various compositional criteria established in the Schedules to the 1997 Regulations.

(e) Instructions as to the preparation, if necessary, and a statement regarding the importance of following the instructions.

Penalties and enforcement

If any person contravenes the labelling or compositional criteria specified in these Regulations, he can be found guilty of an offence and held liable on summary conviction to a fine not exceeding £5,000. Each UK food authority is given powers to enforce and execute these Regulations in its own area. A defence is available where the person charged can prove that the food in respect of which the offence is alleged to have been committed was intended for export to a country which has legislation analogous to these Regulations and it complies with the legislation of the exporting country. Various sections of the Food Safety Act 1990 apply for the purposes of these Regulations and in particular, the defence of due diligence provided for in section 21.

Amendments

The 1997 Regulations have been amended by the Processed Cereal-based Foods and Baby Foods for Infants and Young Children (Amendment) Regulations 1999 (SI 1999 No 275).

These Regulations prohibit the manufacture and sale of processed cereal-based foods and baby foods which contain any specified added nutrient in excess of the maximum limit and therefore it introduces a new Schedule into the overall regulations providing for maximum limits for vitamins, minerals and trace elements if added to these foodstuffs. In relation to compositional requirements for baby foods these Regulations specify that if cheese is mentioned together with another ingredient in the name of a savoury product for babies then there are specific compositional requirements for its inclusion, while specified sauces in sweet dishes are exempt from protein requirements detailed in the Schedules to the 1997 Regulations.

6.6 **Infant and follow-on formulae**

Within the framework of Directive 89/398/EEC the European Union has introduced Council Directive 91/321/EEC[6] on infant formulae and follow-on formulae. **Infant formulae** means foodstuffs intended for use by infants during the first four to six months of life, whereas **follow-on formulae** means foodstuffs intended for particular nutritional use by infants aged over four months and which constitute the principal liquid element in a progressively diversified diet for such infants.

This Directive has been implemented into UK law by the Infant Formulae and Follow-on Formulae Regulations 1995 (SI 1995 No 77). Under regulation 8 the compositional requirements of infant formulae and follow-on formulae are laid down.

Manufacturing requirements

An "**infant formulae**" must be manufactured only from:

(1) The protein sources in food ingredients specified in Schedule 1 to the Regulations.

(2) Other food ingredients the suitability for use by infants from birth has been established by generally accepted scientific data.

(3) The composition of an infant formula should conform to the criteria specified in Schedule 1.

Similarly, under regulation 9 a "**follow-on formulae**" must be manufactured from:

(1) The protein sources in food ingredients specified in Schedule 2.

(2) Other food ingredients the suitability for use by infants aged over four months have been established by generally accepted scientific data.

---

6   OJ 1991 L175/35.

(3) In the manufacture of any such follow-on formula the prohibitions and limitations on the use of food ingredients as set out in Schedule 2 must be observed and the composition of the food product must conform to the criteria specified in Schedule 2 to the Regulations.

Under the Regulations it is specified that if an infant follow-on formulae is not at the time it is sold ready for use, nothing more than the addition of water should be required to make it ready for such use. In respect of vitamins, mineral substances, amino acids and other nitrogen compounds and other substances having a particular nutritional purpose Schedule 3 to the Regulations specifies how these products may be used in the process and manufacture of these baby foods and the requirements of Schedule 3 must be adhered to.

Labelling requirements

"Infant formulae" must be labelled as follows:

- In the case of a product where the protein source is not entirely cow's milk proteins, the name "infant formula" must be used.
- In the case of a product where the protein source is entirely cow's milk proteins, the name "infant milk" must be used.
- A statement to the effect that the product is suitable for infants from birth when they are not breast-fed.
- In the case of a product which does not contain added iron, a statement to the effect that the product does not contain the total iron requirements as recommended for an infant over the age of four months and that this should be made up from additional sources.
- The available energy value expressed in kJ and kcal and the content of proteins, lipids and carbohydrates per 100 mls of the product ready for use.
- The average quantity of each mineral substance and of each vitamin mentioned in Schedule 1 and where applicable of choline, inositol and cartnitine per 100 mls of the product ready for use.
- Instructions for appropriate preparation of the product and a warning against the health hazards of inappropriate preparation.
- The words "an important notice" immediately followed by a statement concerning the superiority of breast-feeding and a statement recommending that the product be used only on the advice of an independent person qualified in medicine, nutrition or pharmacy.

- The labelling of an "infant formulae" must not include any picture of an infant or any other picture or text which may idealise the use of the product.

"Follow-on formulae" must be labelled as follows:

- In the case of a product where the protein source is not entirely cow's milk proteins, the name "follow-on formula" must be used.
- In the case of a product where the protein source is entirely cow's milk proteins, the name "follow-on milk" must be used.
- A statement to the effect that the product is suitable only for use by infants over the age of four months, and that it should form only part of a diversified diet and that it should not be used as a substitute for breast milk during the first four months of life.
- The available energy value expressed in kJ and kcal and the content of proteins, lipids and carbohydrates per 100 mls of the product ready for use.
- The average quantity of each mineral substance and each vitamin mentioned in Schedule 2 of the Regulations and where applicable of cholin, inositol and carnitine per 100 mls of the product ready for use.
- Instructions for appropriate preparation of the product and a warning against the health hazards of inappropriate preparation.
- The labelling must not include information discouraging breast-feeding and must not contain the terms "humanised", "maternalised" or any similar term suggesting that the product is equivalent or superior to breast milk.

Promotion restrictions

Regulations 17 and 18 introduce restrictions on the advertising of "infant formulae" and "follow-on formulae" so that no person will publish or display any advertisement for these foodstuffs except in a publication specialising in baby care or in a scientific publication or in a retail trade publication where the intended readership is other than the general public. These advertisements must contain only information of a scientific and factual nature and must not imply or seek to create a belief that bottle-feeding is equivalent to or superior to breast-feeding.

Regulation 19 introduces restrictions on the promotion of such foodstuffs at the point of sale or the use of free coupons. Finally, regulation 20 stipulates that a manufacturer or distributor of any infant formula must not provide free samples of the product or discounted versions of the product in order to promote the sale of an infant formula either to the general public, pregnant women, mothers or members of the families of persons associated with children either directly or indirectly through the healthcare system or health workers.

Penalties and enforcement

Any person who contravenes or fails to comply with the 1997 Regulations will be guilty of an offence and can be liable on summary conviction to a fine not exceeding £5,000. The Regulations specify that local food authorities must enforce and execute the Regulations in their own area and various provisions of the Food Safety Act 1990 apply for the purposes of these Regulations, in particular, sections 8, 14 and 15 as well as section 21 of the 1990 Act which introduces the defence of due diligence.

Amendments

The 1995 Regulations have since been amended by the Infant Formulae and Follow-on Formulae (Amendment) Regulations 1997 (SI 1997 No 451). These 1997 Regulations introduce amendments in relation to the labelling of infant formula and follow-on formula by making it mandatory for the inclusion of the average quantity of nutrients mentioned in Schedule 3 to the Regulations to be included on the label. Also a new Schedule 8 is introduced to the Regulations, covering the reference values for nutritional labelling for foods intended for infants and young children.

6.7 Irish legislation

Cocoa and chocolate products

Directive 73/241/EEC has been implemented by the European Communities (Cocoa and Chocolate Products) Regulations 1991 (SI 1991 No 180). These Regulations give effect to the EU Directives which prescribe standards for the composition and labelling of cocoa and chocolate. The maximum amount of vegetable fat other than cocoa butter in chocolate products is fixed. Where appropriate, there must be a declaration on the label or adjacent notice that the products contain such fat. Compositional and quality standards are provided in the

Regulations for various chocolate products including plain chocolate, chocolate flakes, milk chocolate, gianduja nut milk chocolate, fill chocolate etc. Permitted additional ingredients are listed and their labelling requirements as to the cocoa and milk solids content.

Enforcement

Trade in products that do not conform to these Regulations is prohibited and authorised officers are granted the power of entry to premises and also the power to take samples in order to ascertain whether the Regulations are being correctly enforced.

Natural mineral waters

Directive 80/777/EEC has been implemented in Ireland by the European Communities (Natural Mineral Waters) Regulations 1986 (SI 1986 No 11). These Regulations simply point out that the various articles in the European Council Directive dealing with the compositional content and labelling, packaging and advertising of natural mineral waters as well as the criteria contained in Annex 1 of the Directive following the introduction of these Regulations are applicable in Irish law. Regulation 9 states that in relation to Article 4(1) of the Council Directive it must not be construed in such a manner as to prohibit in any way the use of natural mineral waters in the manufacture of soft drinks.

Penalties and enforcement

Regulation 10 details the powers of authorised officers to enter premises and take samples in order to ensure that these Regulations are being enforced. A person guilty of an offence under these Regulations shall be liable under summary conviction to a fine not exceeding £1,000 or at the discretion of the court to imprisonment for a term not exceeding six months or to both a fine and imprisonment. In relation to a person guilty of the offence of obstructing an authorised officer to take samples or to look at documentation to ensure that the Regulations are being enforced, the person prosecuted can be liable upon summary conviction to a fine not exceeding £750. There is also provision under regulation 13 for an offence committed by a body corporate to be prosecuted under these Regulations.

Processed cereal-based foods

The European Communities (Processed Cereal Based Foods and Baby Foods for Infants and Young Children) Regulations 1998 (SI 1998 No 241) implements Directive 96/5/EEC on cereal-based baby foods. Under regulation 8 processed cereal-based foods must comply with the compositional criteria specified in Schedule 1 to the Regulations and baby foods the compositional criteria described in Schedule 2. Only the nutritional substances listed in Schedule 4 may be added in the manufacture of either processed cereal-based foods or baby foods. The Minister has powers under regulation 10 to make an order stipulating the maximum levels of any substances included in such foodstuffs or to establish such microbiological criteria he considers appropriate in the manufacture of such foodstuffs.

Penalties and enforcement

The Regulations are enforced and executed by each health board in respect of its functional area and/or officers of the Minister for Health and Children under the conditions laid down in the European Communities (Official Control of Foodstuffs) Regulations 1998 (SI 1998 No 85). Under regulation 15 a person must not manufacture, prepare, import, distribute, market and/or label any product which does not comply with these Regulations and if a person contravenes the Regulations he will be liable on summary conviction to a fine not exceeding £1,000 or at the discretion of the court to imprisonment for a term not exceeding six months or to both. Offences by a body corporate can also be prosecuted under the Regulations and proceedings for an offence under these Regulations must be instituted within 12 months from the date of the offence or any time within 12 months from the date on which knowledge of the commission of the offence came to the attention of an authorised officer. An offence under these Regulations may be prosecuted either by the Minister or the health board within whose functional area the offence was committed.

Infant and follow-on formulae

The European Communities (Infant Formulae and Follow-on Formulae) Regulations 1998 (SI 1998 No 243) implement Directive 91/321/EEC covering these foodstuffs. These Regulations introduce conditions for the marketing of such foodstuffs and the compositional criteria are included in Schedules 1 and 2 to the Regulations. Labelling, advertising and presentation requirements are included in regulations 9 and 10. These Regulations are enforced and executed by each health board in respect of its own functional area in line with the European Communities (Official Control of Foodstuffs) Regulations 1998.

Penalties and enforcement

A person must not manufacture, prepare, import, export, distribute, market, advertise and/or label any product or promotional material which does not comply with these Regulations and if found guilty will be liable on summary conviction to a fine not exceeding £1,000 or at the discretion of the court to imprisonment for a term not exceeding six months or to both. A body corporate may be prosecuted under the Regulations. Proceedings for an offence under these Regulations must be instituted within 12 months from the date of the offence or at least 12 months from the date of knowledge of the commission of the offence. Any offence under these Regulations may be prosecuted either by the Minister for Health or a health board within whose functional area the offence was committed.

6.7 **Conclusions**

This chapter has covered a number of diverse areas of UK food law, including food products such as chocolate, natural mineral waters and special nutritional foods like baby foods. What all this legislation has in common is that it is "compositional" by nature, covering in some detail what these products must consist of if they are to be manufactured and marketed in both the United Kingdom and the European Union. There are many other similar pieces of UK law covering different food products. This type of law is no longer a preferred form of legislation in the European Union. Following the *Cassis de Dijon*[7] case there has been a tendency in the EU to prefer enacting less descriptive and compositional food legislation. *Cassis de Dijon* introduced the "mutual recognition" principle whereby if a food product was accepted to be marketed in one Member State then it should be permitted to be marketed throughout the internal market. With the future enlargement

---

7   [1979] ECR 649.

of the EU to include a number of countries from Central and Eastern Europe, the "mutual recognition" principle is going to become ever more important.

The one area where there is likely to be more legislation of a compositional nature is in relation to what are known as "functional foods". There is no simple definition for a functional food because of the diverse nature of the many food products that can be fortified with functional ingredients. These types of food products generally include many types of healthy foods that have additional ingredients, which make them "healthy" according to the manufacturers. Examples would include special yoghurts and milk shakes. Unlike food additives, these fortified ingredients do make a substantial difference to the nature of the food product which the consumer purchases. A number of such products are on the market at present, but with technological advances it seems likely that this is an area of food law which will need further regulation in the future. As the functional ingredients are added to enhance the food product, it would seem that the only way to legislate for such products would be in a "compositional" manner, specifying the percentage of fortified ingredients that could be added to different food products.

Chapter 7

· **Food Additives** ·

Summary

- EU Framework Directive 89/107/EEC covering food additives
- Food additives defined – authorisation process – perceived need – manufacturing process
- Specific legislation for different types of additives
- Sweeteners in Foods Regulations 1995
- Colours in Foods Regulations 1995
- Miscellaneous Food Additives Regulations 1995
- Food additives law in Ireland

7.1 **Introduction**

Council Directive 89/107/EEC[1] contains the basic legal provisions concerning food additives in the European Union. It is sometimes referred to as the "Framework Directive" and subsequent specific rules concerning colours, sweeteners and miscellaneous additives have been made under its auspices. The framework Directive defines a food additive as meaning a substance not normally consumed as a food in itself and not normally used as an ingredient in a foodstuff for its nutritive value. Instead, when it is used in producing a particular food product, it becomes directly or indirectly a component of such foods. The inclusion of an additive in one of the 24 categories in Annex 1 of the Directive is based on the principal function formally associated with that food additive, but that will not exclude the additive being authorised to have a number of functions thereby being listed in a number of the categories in Annex 1. In that case, a food additive could function as a sweetener, preservative and glazing agent all at the same time and would therefore be listed in three of the categories in Annex 1. Food additives are included in a list in Annex 1 on the basis

---

1   OJ 1989 L40/27.

of the general criteria on additives which are described in Annex 2 of the Directive.

The Directive clearly states that food additives can only be authorised provided that:

- A reasonable technological need for their use can be demonstrated which cannot be achieved by any other means.
- They present no hazard to human health at the level of use proposed based on existing scientific evidence.
- They do not mislead the consumer.

These provisions are elaborated further especially in terms of the criteria of "need" which is introduced into the Directive. This means that only where it can be shown that the use of the additive provides demonstrable advantages for the consumer and serves one of four different purposes mentioned below can it be proved that there is "need" for the use of that particular additive in the manufacture of a specific foodstuff. The use of the food additive should serve one of the following functions:

(1) To preserve the nutritional quality of the food.

(2) To provide necessary ingredients or constituents for food products manufactured for groups of consumers having special dietary needs.

(3) To enhance storage aspects of a food product.

(4) To assist in the manufacture, processing and preparation of food products, so long as the additive does not disguise the effects of the use of sub-standard raw materials or unhygienic practices.

All food additives must be kept under continuous observation and must be re-evaluated whenever necessary in the light of the changing conditions of use and new scientific information. The Directive introduced specific labelling requirements for additives which were implemented into UK law by the Food Additives Labelling Regulations 1992 (SI 1992 No 1978, discussed in Chap 6). Those Regulations not only implemented the labelling requirements of Council Directive 89/107/EEC, but also defined food additives and listed in Schedule 1 the categories of permitted food additives in the United Kingdom. A number of specific types of additives, *e.g.* sweeteners, colours and miscellaneous additives have been considered in separate EU Directives taking into account the overall provisions of the framework Additives Directive 89/107. When implemented into UK law, these Directives now form the main body of UK legislation pertinent to the whole area of food additives.

7.2 **Sweeteners**

The Sweeteners in Food Regulations 1995 (SI 1995 No 3123) implement European Parliament and Council Directive 94/35/EEC[2] on sweeteners for use in foodstuffs. The Regulations define a sweetener as meaning any food additive which is used or intended to be used:

- to impart a sweet taste to food; or
- as a table top sweetener.

The Regulations prohibit:

(1) The sale of unauthorised sweeteners to the ultimate consumer or for use in or on food.

(2) The use of sweeteners in food except those authorised and used as prescribed in the Regulations.

(3) The sale of food that contains unauthorised sweeteners or that contains permitted sweeteners unless these are used as prescribed by the Regulations.

(4) The use of sweeteners in food specially prepared for infants and young children as specified in Council Directive 89/398/EEC[3] relating to foodstuffs intended for particular nutritional uses.

The 1995 Regulations contain a Schedule which lists permitted sweeteners and the foods in or on which they may be used. This Schedule has been subsequently updated by the Sweeteners in Food (Amendment) Regulations 1997 (SI 1997 No 814). The following are some examples from the Schedule:

Foodstuff	*Permitted sweetener*	*Maximum useable dose*
Cereal based deserts	E953 Isomalt	*Quantum satis*
Confectionery, chewing gum, edible ices	E957 Thaumatin	50 mgs/kgs
Food supplements	E957 Thaumatin	400 mgs/kgs

---

2   OJ 1994 L237/3.
3   OJ 1989 L186/27.

These examples show the type of information provided in Schedule 1 to the Regulations on permitted sweeteners. Those permitted sweeteners where the maximum usable dose is referred to as *quantum satis* are cases where no maximum levels are prescribed, but the permitted sweetener may be used in the food product in accordance with good manufacturing practice, provided the level does not exceed that necessary to achieve the intended purpose and that such use does not mislead the consumer. The example of thaumatin demonstrates that the maximum usable dose can change for different foodstuffs. In the case of well known sweeteners, such as aspartame, syclamic acid and saccharin there is a long list of foodstuffs where these particular sweeteners are permitted and similarly a variety of different maximum usable dosages concerning that particular sweetener.

Penalties

If any person contravenes or fails to comply with any of the provisions of the regulations and in particular Schedule 1, he will be guilty of an offence and liable on summary conviction to a fine not exceeding £5,000. In any proceedings for an offence under the Regulations it will be a defence for the person charged to prove that the food or, as the case may be, sweetener in respect of which the offence is alleged to have been committed was intended for export to a country which has legislation analogous to these Regulations and that such food or sweetener complies with that legislation. This defence is only available in the case of a country not a member of the European Union. Regulation 9 of the 1995 Regulations specifies that sections 8, 14 and 15 of the Food Safety Act 1990 shall apply for the purposes of the 1995 Regulations and that in any prosecution taken under the Regulations, various other sections of the Food Safety Act will come into play, including section 21 concerning the defence of due diligence.

Amendments

The 1995 Regulations have most recently been amended by the Sweeteners in Food (Amendment) Regulations 1999 (SI 1999 No 982). These new Regulations bring the Sweeteners in Food Regulations 1995 up to date in relation to a number of technical matters.

7.3 Colours for use in foodstuffs

The Colours in Food Regulations 1995 (SI 1995 No 3124) adopt and implement EP and Council Directive 94/36/EEC[4] concerning colours for use in foodstuffs and apply all the lists contained in the Annexes to that Directive. The Regulations define a "colour" as meaning a food additive which is used or is intended to be used for the primary purpose of adding or restoring colouring in a foodstuff. It includes:

- any natural constituent of food and any natural source not normally consumed as food as such and not normally used as a characteristic ingredient of food;
- any preparation obtained from food or any other natural source material by physical and/or chemical extraction that results in selective extraction of the pigment relative to the nutritive or aromatic constituent.

Under regulation 3 of the 1995 Regulations it is an offence for a person to use in or on any food any colour other than a permitted colour as laid down in the various Schedules to the 1995 Regulations. Other provisions of the Regulations specify those colours to be used for the purpose of health marking of carcasses under the Fresh Meat Regulations 1995 and also the colours which should be used in relation to the stamping of egg shells under the 1991 Regulations on marketing standards for eggs.

The five Schedules to the Regulations consist of a list of:

- the permitted colours;
- food that generally may not be coloured;
- foods that may only contain certain colours;
- colours that may only be used in certain foods; and
- two further lists of colours that may be used more freely in foodstuffs, although specific restrictions apply to some of them.

Schedule 1 to the 1995 Regulations lists 43 permitted colours from E100 curcumin to E180 litholrubine BK. Schedule 2 lists 33 different types of foodstuffs which may not contain added colours except where specifically provided in Schedules 3, 4 or 5. These foodstuffs

---

4   OJ 1994 L237/13.

include bottled water, eggs and egg products, pasta, tomato based sauces etc. Schedule 3 contains a lengthy list of foods in which the use of colour is restricted, in some cases colours that are allowed may only be used to maximum specified quantities. The list includes certain cheeses, spirit drinks and other alcoholic drinks, jams, jellies and marmalades and several meat, vegetable and cereal products. The following are examples:

Foodstuff	*Permitted colour*	*Maximum level*
Luncheon meat	E129 allura red	25 mgs/kgs
Dried potato granules and flakes	E100 curcumin	*quantum satis*
Fruit flavour breakfast cereals	E120 cochineal E162 beetroot red E163 anthochanins	200 mgs/kgs (individually or in combination.

Schedule 4 covers colours permitted for use in certain foodstuffs only. Examples of these are:

Foodstuff	*Permitted colour*	*Maximum level*
Kippers	E154 brown FK	200 mgs/kgs
Edible cheese rind	E180 litholrubine BK	*quantum satis*

Schedule 5 consists of further permitted colours for use in foodstuffs and contains a list of colours whose use is more flexible although certain important restrictions are prescribed. Most of the foods in which they are allowed are widely used and commonly manufactured. The Schedule is divided into three parts:

- Part 1 lists 15 particular colours that may be used at quantum satis level from E101 riboflavin to E172 iron oxides and hydro oxides;
- Part 2 contains a list of 18 colours from E100 curcumin to E161 b lutein that may be used singularly or in combination in the foods listed in Part 3;
- Part 3 contains a list of foodstuffs in which the colours listed in Parts 1 and 2 are permitted and the maximum level permitted for such colours.

Food	Maximum level
Non-alcoholic flavoured drinks	100 mgs/litre
Flavoured processed cheese	100 mgs/kilo
Mustard	300 mgs/kilo

Whichever permitted colours are used in the food products listed above if they are a colour as listed in Part 1 of Schedule 5 they will have to be used at a level of *quantum satis* and if they are from Part 2 of Schedule 5 they will have to be used at a level as specified in the table above.

Penalties

Under regulation 9, if any person contravenes or fails to comply with any of the provisions of the Regulations or its Schedules, he will be guilty of an offence and liable on summary conviction to a fine not exceeding £5,000. In any proceedings for an offence under these Regulations it will be a defence for the person charged to prove that the food or, as the case may be, the colour in respect of which the offence is alleged to have been committed was intended for export to a country outside the European Union which has legislation analogous to these Regulations and that the foodstuff or colour complies with that legislation. For the purposes of the Regulations sections 8, 14 and 15 of the Food Safety Act 1990, apply and in that context various other sections of the 1990 Act come into effect, in particular, section 21 relating to the defence of due diligence.

7.4 Miscellaneous food additives

The Miscellaneous Food Additives Regulations 1995 (SI 1995 No 3187) implement EP and Council Directive 95/2/EEC[5] concerning food additives other than colours and sweeteners. Under regulation 3 no person shall use in or on any food any miscellaneous additive other than a permitted miscellaneous additive as specified in these Regulations and in particular in the Schedules to the 1995 Regulations.

Schedule 1 to the Regulations lists 106 miscellaneous additives which can generally be permitted for use in particular foodstuffs which are

---

2   OJ 1995 L61/1.

specified in a number of later Schedules. These miscellaneous additives range from E170 calcium carbonates to E1450 starch sodium octenyl succinate. Schedule 2 deals with conditionally permitted preservatives and antioxidants. Schedule 3 contains a list of other additives that have miscellaneous functions, but these are restricted to use in certain foodstuffs and generally maximum levels of use are prescribed. The following examples are:

Foodstuff	*Permitted additive*	*Maximum level*
Non-alcoholic flavoured drinks	E444 sucrose acetate isobutyrate	300 mgs/litre
Dried egg white	E1505 triethyl citrate	*quantum satis*

Schedule 4 to the Regulations lists permitted carrier and carrier solvents and Schedule 5 lists the purity criteria for miscellaneous additives. Schedule 6 to the Regulations importantly lists those foodstuffs which even though a miscellaneous additive may be listed in Schedule 1 are generally prohibited from use in these foodstuffs. These foodstuffs include honey, butter, pasteurised and sterilised milk and cream, coffee, dried pasta etc. Schedule 7 lists those foodstuffs in which a limited number of miscellaneous additives listed in Schedule 1 may be used. Examples are:

Foodstuffs	*Permitted additive*	*Maximum level*
Mozzarella cheese	E270 lactic acid E330 citric acid E575 glucono-delta-lactone	*quantum satis*
Pineapple juice	E296 malic acid	3 grms/litre
Fruit juices	E330 citric acid	3 gms/litre

Finally, Schedule 8 lists miscellaneous additives that are permitted in infant formulae, follow-on formulae and weaning foods used for infants and young children.

Penalties

Under regulation 7 any person who contravenes or fails to comply with any of the provisions of the Regulations or those specifications noted in its Schedules will be guilty of an offence and liable on summary conviction to a fine not exceeding £5,000. As with the previous Regulations concerning colours and sweeteners, it is a defence for a person charged to prove that the food, or as the case may be food additive, in respect of which the offence is alleged to be committed was intended for export to a country outside the European Union whose legislation is analogous to these Regulations and that the food or food additive concerned complies with that legislation. Similarly, sections 8, 14 and 15 of the Food Safety Act 1990, apply for the purposes of these Regulations and various other sections of the 1990 Act thereby can be construed for the purposes of these Regulations in any court proceedings. Most importantly, therefore, the section 21 defence of due diligence is permitted under these Regulations.

Amendments

The 1995 Regulations have been amended on two subsequent occasions.

(1) The Miscellaneous Food Additives (Amendment) Regulations 1997 (SI 1997 No 1413) which implement Directive 96/85/EEC by adding processed eucheuma seaweed to the list of miscellaneous additives generally permitted for use in foods. Other technical amendments were made to the 1995 Regulations bringing it into line with the new EU legislation.

(2) The Miscellaneous Food Additives (Amendment) Regulations 1999 (SI 1999 No 1136) provide for a number of new miscellaneous additives to be added to the various Schedules listed in the 1995 Regulations and also introduce a number of additional changes to the 1995 Regulations in terms of technical progress.

7.5 Food additives legislation in Ireland

There are two main pieces of food legislation covering the whole area of additives on the Irish statute books. The first of these is the European Communities (General Provisions on the Control of Additives and in particular Colours and Sweeteners for Use in Foodstuffs) Regulations 1995 (SI 1995 No 344). These Regulations implement the main provisions of Directives 89/107/EEC concerning food additives, 94/35/EEC on sweeteners and 94/36/EEC on colours.

These Regulations lay down the general provisions for the use of food additives including sweeteners and colours in food products in Ireland. Only those additives contained within the various Schedules to the Regulations may be used in the food manufacturing process.

Schedule 1 to the Regulations lists various categories of food additives that are permitted to be used and specifies the maximum useable level of such additives in various foodstuffs. Interestingly, regulation 6 provides that where the Minister for Health is of the view that the use or intended use in foodstuffs of any food additive, although complying with these Regulations, endangers human health, he may take appropriate measures including the temporary suspension or restriction of trade in that foodstuff or food additive in order to protect public health.

Other provisions of the Regulations lay down the specific labelling requirements for such additives when used in the food manufacturing process. In most cases, the name of the food additive as specified in the various Schedules to the Regulations must be placed on the food package, as well as the EEC number and in the absence of such information at least a description of the product/additive that is sufficiently precise to enable it be distinguished from products with which it could be confused. Any other necessary information in terms of special conditions of storage, directions for use, a mark identifying the batch or lot and the business name and address of the manufacturer, packer or seller established within the EU must be included in order to provide as much information as possible to the ultimate consumer.

Those sweeteners which can be used as a food additive in the food manufacturing process are listed in Schedule 2 to the Regulations and the Regulations also specify those foods in which the term "with no added sugar" and "energy reduced" can be used. Colours that may be used as a food additive are listed in Schedule 4, Part A to the Regulations and Parts B, C, D and E of the same Schedule lists the various maximum dosages for such food additives in a food product.

These Regulations are enforced and executed by each health board in Ireland. An authorised officer of the health board is given powers to enter premises and seize any foodstuff or food additive in pursuance of the provisions of these Regulations which he believes is in contravention of the regulations. The authorised officer can also apply to a judge of the district court for an order directing that a foodstuff or food additive be destroyed or otherwise disposed of because it is unfit for human consumption. The district court will decide whether the foodstuff or food additive complies with the particular Regulations. In any such proceedings it will be a defence for the person charged to show that the food in respect of which the offence is alleged to have been committed was intended for export and complies with the domestic food legislation of the importing country.

A person guilty of an offence under the 1995 Regulations shall be liable on summary conviction to a fine not exceeding £1,000 or at the discretion of the court to imprisonment for a term not exceeding six months or to both. Provisions are contained within the Regulations that a body corporate may be prosecuted under the Regulations and any offence under the Regulations must be instituted within 12 months from the date of the offence. Any offence under these Regulations will be prosecuted either by the Minister for Health or by a health board in whose functional area the offence was committed.

The European Communities (Food Additives other than Colours and Sweeteners) Regulations 1999 (SI 1999 No 288) give effect to Directive 95/2/EEC on miscellaneous additives. As with the previous Regulations those additives permitted for use in foodstuffs other than colours and sweeteners are listed in Schedules 1, 3, 4 and 5 to the Regulations. These Schedules also mention the food products and the maximum level in which such additives can be used in terms of the food manufacturing process.

The health boards are given powers to ensure the enforcement of these Regulations and can send authorised officers into food premises to ensure that the Regulations are being applied correctly. Any person who trades or uses products, which do not comply with these regulations, will be guilty of an offence and liable on summary conviction to a fine not exceeding £1,500. A body corporate may also be prosecuted under the regulations and the Minister of Health or health board will prosecute offences under these Regulations.

7.6 Conclusions

With the trend for more convenience and processed foods, the market for the production of food additives will expand. At the same time, with increased consumer concern over the safety of food products, there is further demand to use fewer additives in foods. Healthy eating has influenced the development of fat replacers and new sweeteners categories in terms of acceptable food additives. In that case, there is likely to be new legislation adapting the existing UK law on food additives to these new technical developments.

However, at EU level, especially in the European Parliament, there is considerable resistance to allowing the various lists, which have been implemented into UK law, to be extended. Indeed, the European Parliament would advocate that the additive lists should be shortened. As this debate is likely to continue, if the UK government wants to advocate the inclusion of a "new" additive to these lists, it will have a very difficult battle on its hands to persuade its EU counterparts. In that case, the existing body of UK law on food additives is likely to remain much the same for the foreseeable future. Any major changes to these additive lists in the future are most likely to advocate the exclusion of certain additives, which after rigorous scientific tests have been found to pose a threat to consumer health.

Chapter 8

· **Food Standards Agency** ·

Summary

- White Paper on a Food Standards Agency (January 1998) and subsequent draft Food Standards Bill 1998
- Food Standards Act 1999
- New agency's main objective – "to protect public health"
- Food Standards Agency (FSA) has a "duty" to provide advice to consumers, local authorities, ministers and the public
- FSA has powers to undertake investigations into food safety problems at any point in the food chain
- FSA will monitor the performance of local authorities in relation to food safety
- FSA will have a major role in relation to animal feeding stuffs, licensing of medicines including veterinary medicines, pesticides and genetically modified organisms (GMOs)
- Role and powers of Food Safety Authority of Ireland

8.1 **Introduction**

In the aftermath of the BSE crisis, both at national and European level, there has been much discussion about how independent food agencies could alleviate many of the enforcement problems faced in this crisis while at the same time increasing consumer confidence in the food they purchase. In the United Kingdom, the Labour Party in opposition commissioned Professor Philip James to prepare a report on the functions and structures of a Food Standards Agency. Professor James presented his blueprint for a Food Standards Agency to the incoming Prime Minister, Tony Blair, two days after his election victory in May 1997. Subsequently a White Paper on a Food Standards Agency was published in January 1988 which advocated the establishment of the following:

"A powerful body which will be responsible for protecting public health by promoting a safer food supply and ensuring that consumers have the information they need to be able to choose a healthy diet."

The essential issues such an agency was expected to resolve were:

- The potential for conflicts of interest within the Ministry of Agriculture and Food arising from its dual responsibility for protecting public health and for assisting and maintaining dynamic and viable agriculture/food industries.
- The fragmentation and lack of co-ordination between the various government bodies involved in food safety.
- The uneven enforcement of food law.

A draft Food Standards Bill was published in January 1999 which was the subject of pre-legislative scrutiny by a select committee of the House of Commons, whose report was published in March 1999. This Bill to establish a Food Standards Agency received royal assent on 11 November 1999. When established the Food Standards Agency will be a UK body accountable to Parliament, the Scottish Parliament, the National Assembly for Wales and the Northern Ireland Assembly. It will operate at arms length from ministers under the day-to-day responsibility of a chairman, deputy chairman and board members. The intention is that the Food Standards Agency should become operational in Spring 2000.

8.2 **Food Standards Act 1999**

Introduction

The main purpose of the Act is to establish the Food Standards Agency, provide it with functions and powers and to transfer to it certain functions in relation to food safety and standards under other UK Acts. It sets out the agency's main objective for protecting public health in relation to food, the functions that it will assume in pursuit of that aim and gives the agency the powers necessary to enable it to act in the consumer's interest at any stage in the food production and supply chain. The Act provides for the agency's main organisational and accountability arrangements. In addition, it provides powers to establish a scheme for the notification of the results of tests of food borne diseases.

With the enactment of the Food Safety Act 1990, it was felt that the UK government had introduced a major reforming piece of legislation aimed at protecting consumer health in the whole area of food safety. The ensuing controversies over Salmonella in eggs and the BSE crisis have undermined consumers' confidence in the food they purchase. It therefore seems opportune that at the end of the decade the Food Standards Act has been enacted which will not only establish an independent agency in the guise of the Food Standards Agency, but also attempt to fill in many of the gaps and loopholes which have been seen to exist in relation to the enforcement and implementation of both UK and European food law in the United Kingdom. In the years ahead it is likely that the establishment of the Food Standards Agency will be seen to have been one of the most far-sighted and reforming pieces of legislation introduced by the Blair government.

Objective and role of Food Standards Agency

The Food Standards Act 1999, consists of 43 sections and six Schedules. In section 1(2) it is stated quite clearly that the main objective of the establishment of a Food Standards Agency is:

> "To protect public health from risks which may arise in connection with the consumption of food (including risks caused by the way in which it is produced or supplied) and otherwise to protect the interests of consumers in relation to food."

Food safety therefore is central to the establishment of this agency, but the role of the Food Standards Agency will also cover issues such as nutrition and diet and protecting the wider food-related interests of consumers, sometimes referred to as food standards. In particular, this will cover such matters as the labelling and composition of food products. It is important to note that the way in which the Food Standards Agency carries out these functions is limited by the requirements set down in subsequent provisions of the Act, in particular sections 22 and 23. These sections provide for the agency to carry out its functions in accordance with a statement of objectives and practices that have been approved by the appropriate authorities, *e.g.* local authorities, meat hygiene service etc. Section 22 requires the agency immediately on its establishment to prepare and publish a statement of its general objectives and practices. Among various general objectives to be included in this statement, section 22(2) specifies that the agency must address three in particular. These relate to:

(1) Consulting with interested parties on the agency's activities.

(2) Facilitating proper consultation between the agency and other government departments, local authorities and other public authorities on matters of mutual interest.

(3) Ensuring that the agency's activities and decisions are open and transparent to the public.

In carrying out any of its functions, section 23 specifies that the agency must have due regard to its statement of general objectives and practices and take account of relevant advice from advisory committees. In its decision-making process the agency is required to take account of the nature and magnitude of the risks which the action under consideration is designed to address. A risk to health is highlighted as of particular importance but other risks in relation to consumer protection (*e.g.* where labelling may mislead consumers) may also be relevant. The agency is also required to take account of any uncertainty in the evidence, *e.g.* where it is taking decisions in relation to a risk which is potentially very serious, but about which there is very little evidence, the agency is likely to want to take a precautionary approach. Also, the likely costs and benefits associated with any course of action under consideration should be taken into account in the decision-making process. The agency must therefore balance obvious compliance costs as well as matters such as restriction of consumer choice, against the benefits of reduced risks to public health etc arising from any of its actions. These two sections of the Act give a taste of the government's philosophy behind the establishment of the Food Standards Agency, as although the agency has been given considerable new powers in the whole area of food safety, the government is obviously concerned that the agency must be able to defend rigorously any actions or decisions it takes and show that they have taken all due account of the scientific evidence, advice from government advisory committees and looked at the risks and costs involved in the outcome of any action or decision by the Food Standards Agency.

Composition of Food Standards Agency

The Food Standards Act provides for the agency to have a chairman, deputy chairman and eight to 12 other members of whom one will be appointed by the National Assembly for Wales, two by the Scottish Ministers and one by the Department of Health and Social Services for Northern Ireland. The rest of the members of the Food Standards Agency Board will be appointed by the Secretary of State for Health.

The Act requires that the members will not be appointed to be representative of any particular interest or industrial sector. The Act requires the appropriate authorities to consider whether the person's financial or other interests, *e.g.* shares in a major food manufacturer, are likely to compromise his position as a member of the agency. A chief executive will be appointed to run the agency and his responsibility will be to ensure that the agency is run efficiently and effectively. He will be responsible to the agency's members for the day-to-day running of the agency itself and also be the agency's accounting officer.

Directors shall also be appointed for Wales, Scotland and Northern Ireland, each of whom will be responsible under the aegis of the chief executive for securing that the activities of the agency in Wales, Scotland or Northern Ireland are carried out. In the same vein advisory committees will be established for Wales, Scotland and Northern Ireland for the purposes of giving advice or information to the agency about matters connected with its functions in those parts of the United Kingdom. Likewise, the agency has powers to establish other specialist advisory committees if it so wishes. This issue is dealt with in greater detail in Schedule 2 to the Act where it is also noted that a number of existing non-statutory advisory committees will fall under the remit of the agency. These include the Advisory Committee on the Microbiological Safety of Food (ACMSF), the Advisory Committee on Novel Foods and Processes (ACNFP) and the Food Advisory Committee (FAC) and the consumer panel. The schedule also provides that the agency may establish joint committees with another authority or ministerial department. Current examples of joint committees include the Committee on the Medical Aspect of Foods and Nutrition Policy (COMA) and the Advisory Committee on Animal Feedingstuffs.

Development of food policy and advisory role

The main functions of the Food Standards Agency with relation to local authorities, ministers and the public in general are detailed in sections 6 and 7 of the Act. The agency will have the function of providing advice, information and assistance on matters relating to the development of policy on food safety and related areas to any public authority as well as ministers and government departments. The agency will have "a duty to provide" such advice, information and assistance on request unless it is not reasonably practical to do so due to financial costs. The advice or information given could include making recommendations to ministers on the need for new UK food legislation or proposing and drafting secondary legislation in order to improve food safety and standards. Another important aspect of the agency's function in assisting ministers will be to represent the United Kingdom at official level in relevant EU and other international forums.

It is intended that the agency as a UK body will be the primary source of policy advice in relation to food safety and associated food issues to the government as a whole and to the devolved authorities in Wales, Scotland and Northern Ireland. Most of the relevant expertise available to the government and to those national authorities in the area of food safety and standards will therefore reside within the agency and will not be duplicated within other government departments. The agency will also be able to advise on the development of policies by other government departments on matters that are relevant to the agency's own areas of responsibility, *e.g.* advice to the Ministry of Agriculture, Fisheries and Food on activities in relation to farming methods which may have an impact on food hygiene or on relevant consumer protection matters in relation to the Department of Trade and Industry. In relation to any information or advice the agency can give to the general public as whole or to particular representatives of the food industry, the agency will be able for example to:

- run information campaigns on issues of current interest or importance;
- publish scientific data arising from research or surveillance and advise on its interpretation;
- publish information on enforcement activities, such as the existing BSE meat enforcement publications;
- produce leaflets on food hygiene, labelling etc;
- run a consumer help line;
- issue advice for people with food allergies;
- pass on information about developments in food science to the public as a whole and in particular groups such as food producers;
- produce guidance on food safety matters for the food industry;
- issue food hazard warnings, alerting the public to particular problems.

Powers to acquire information

In order to assist the agency to keep abreast of new developments in discharging these general functions, the agency can develop its own scientific know-how by undertaking, commissioning or co-ordinating research. In that case, current research and development projects within the remit of the agency funded by the Ministry of Agriculture, Fisheries and Food and any relevant research funded by other departments, such as the Department of Health, will be transferred to the Food Standards Agency at its inauguration. Another important point to note is that the agency will have the same functions of giving policy advice and information both to public bodies and to the consumer in general in relation to the subject of animal feedingstuffs. The Ministry of Agriculture, Fisheries and Food will retain certain particular responsibilities in relation to animal feedingstuffs, but in future it will be necessary for both the Ministry and the agency to co-operate closely on these matters in order to ensure the safety of consumers.

The Food Standards Act gives the agency powers under sections 10 and 11 to help it fulfill its general functions in obtaining and keeping under review any information relevant to its work in the area of food safety. In order to obtain such information the agency is given powers (under s 10)to undertake surveillance programmes or other such investigations at any point in the food production and supply chain and anywhere else where there might be implications for food safety and related matters, *e.g.* the agency will be able to undertake observations on farms. These specific powers give the agency the opportunity to obtain information either directly or through an authorised person acting on its behalf. These powers therefore replace previous more limited powers contained in section 25 of the Food Safety Act 1990. These new provisions expand the previous powers in the Food Safety Act 1990, to allow the agency to carry out its proposed role in monitoring activities at earlier stages of food production and without the need for further secondary legislation, such as a ministerial order which was the case in the previous 1990 Act. Examples of the types of observations that the agency might carry out are surveillance programmes to investigate the presence of pathogens that could carry risks for human health or of a particular contaminant such as lead in certain types of foodstuffs or surveys of hygiene practices in a certain type of food business, like slaughter houses.

It must be noted that the powers in these sections relate to the gathering of information of a general and representative nature and not to the investigation of individual complaints or failures for which the enforcement powers in the Food Safety Act 1990, will continue to be used by the different enforcement authorities in the United Kingdom. Since the observations made under these sections are not intended for enforcement purposes there is no requirement that these investigations be used to gather evidence in accordance with the kind of safeguards contained in the Police and Criminal Evidence Act 1984, and thus any information obtained could not in general be used directly for the purposes of food law enforcement. Where apparent problems are identified in the course of a surveillance exercise the information gathered would normally be passed to the relevant enforcement authorities who would then take a decision on the need for further investigation.

Powers of entry

In the specific case of section 11 there is provision for an authorised person on the agency's behalf to be given powers of entry in order to carry out a specific surveillance programme in line with the intentions of the agency. An authorised person may, if it appears necessary for him to do so, enter any premises at a reasonable hour, take samples of any articles or substances found on the premises, take samples from any food source found on the premises as well as inspect and copy any records or require any person carrying on such a business to provide him with such facilities, such as records or information or such other assistance as he may reasonably request. At all times the authorised person should on request produce his authorisation from the Food Standards Agency permitting him to carry out this investigation before exercising any powers under section 11.

Offences

Section 11 introduces two new offences:

(1) The authorised person is not permitted to disclose to any other person information such as trade, business secrets which he has obtained while completing the investigation.

(2) Those who intentionally obstruct an authorised officer exercising his powers under this section or fail to furnish information to the officer are also guilty of an offence under section 11.

In both cases the guilty party can be held liable on summary conviction to a fine not exceeding £5,000.

These new powers given to the Food Standards Agency are very much in line with the Directive on Official Control of Foodstuffs.[1]

---

1 Council Directive 89/397/EEC, OJ 1989 L186/23.

Under the Directive general principals regarding the official control of foodstuffs are laid down for each Member State. Under the Directive control measures should comprise one or more of the following operations:

- inspection;
- sampling and analysis;
- inspection of staff hygiene;
- examination of written and documentary material;
- examination of any verification systems established by the commercial undertaking and of the roles obtained.

In the optimum situation the European Union has specified that it would be best if Member States could introduce these control measures in a preventative fashion rather than using them in emergencies. Taking this into account there is little doubt that sections 10 and 11 of the Food Standards Act aim to introduce a far more preventative form of official control in the United Kingdom in the light of the experience of Salmonella outbreaks and the BSE crisis.

Monitoring enforcement

Under the 1999 Act the Food Standards Agency has also been given considerable powers in relation to monitoring the performance of local enforcement authorities in their work in implementing UK and European food law. Under section 12 of the Act the agency is given powers to monitor, set standards for and audit the performance of enforcement authorities *(e.g.* London boroughs, non-metropolitan county councils, borough councils, district or county councils and other local authorities). Powers are given to the agency to request information relating to the enforcement action of local food authorities and the powers of entry for monitoring such enforcement action are contained in sections 13 and 14. These sections provide the specific powers necessary for the agency or an authorised person to carry out the monitoring role provided for in section 12.

Section 41 of the Food Safety Act 1990 makes provision for ministers to require reports and returns, but does not allow for audit visits or the provision of detailed returns, statistics and supporting documentation nor does it provide for ministers to set performance targets in relation to enforcement. In that case the majority of the powers contained in sections 13 and 14 of the Food Standards Act 1999 are therefore new powers in the whole area of the official control of foodstuffs. Similar to section 11 of the 1999 Act, section 14 of the 1999 Act provides for the agency to authorise the use of powers of entry in connection with its enforcement monitoring function. This authorisation may include limitations including requiring any authorised person entering premises to follow various necessary food safety precautions. In a similar vein to section 11, section 14 gives the authorised officer powers to enter and inspect individual premises, the possibility of access to records and data held by the enforcement authority or anyone acting on its behalf and also provides for the taking of samples.

The types of premises that may be entered are specified in section 14(5), and these would include any laboratories that provide the local food authority with services relevant to its enforcement activity (this would include both in-house and independent laboratories). The agency is not expected to publish the details of the performance of the laboratories themselves but the details of the performance of these laboratories will be taken into account in their overall audit of the particular local food authority under investigation. A third category of premises subject to these powers of entry covers any in which food law enforcement can be carried out (*i.e.* food shops, food manufacturers, slaughter houses etc). In section 14 there is provision for an authorised person when entering premises to be accompanied by another person such as an official of the European Commission engaged in a routine audit of Member States enforcement of the provisions of EU food law. These powers are similar to those included in the Additional Measures Concerning Official Control of Foodstuffs Directive 93/99/EEC.[2] That Directive provided for the European Commission to appoint and designate specific officials to co-operate with the competent authorities in the Member States to monitor and evaluate the equivalence and effectiveness of official control systems operated by the competent authorities of the Member States. In that case section 14 of the 1999 Act provides for such officials to assist the Food Standards Authority in the enforcement of the official control of foodstuffs in the United Kingdom.

New offences and penalties

A new offence is also introduced by section 16 of the Food Standards Act 1999 whereby a person who intentionally obstructs an authorised

---

2   OJ 1993 L290/14.

person exercising powers under sections 13 and 14 of the 1999 Act or fails to comply with any requirement imposed under those sections or furnishes information which he knows to be false or misleading when requested to do so by an authorised officer is guilty of an offence. A person guilty of an offence under this section can be liable on summary conviction to a fine not exceeding £5,000.

Emergency orders

The Food Standards Agency is to be given important powers to make "emergency orders" under section 17 of the 1999 Act. Under these provisions the agency will be authorised to exercise on behalf of the Secretary of State for Health powers to make emergency orders under:

(1) Sections 1 and 2 of the Food and Environment Protection Act 1989 (emergency orders).

(2) Section 13 of the Food Safety Act 1990 (emergency control orders).

The Food and Environment Protection Act 1989 deals specifically with pesticides and therefore the agency would be involved in introducing an emergency order if there was a particular food safety problem involving pesticides. On the other hand, the Food Safety Act 1990, covers various food safety issues, in that case the agency has now been given considerable new powers both to introduce these emergency control orders, issue codes of practice and give directions for their enactment by enforcement authorities in particular food emergencies. Under section 17 the Secretary of State will retain powers to introduce such emergency orders in situations which pose a threat to public health in relation to food, but the provisions in section 17 give the agency powers to make emergency orders itself on his behalf. This power does not give the agency the ability to make legislation itself in other areas and in practice it is envisaged that the Food Standards Agency will only make orders in emergency situations where the Secretary of State is not available.

In a similar vein, under section 18 of the 1999 Act, it is specified in Schedule 3 that various functions under different Acts of Parliament are now to be transferred to the Food Standards Agency. The new agency will now have an input into any decision regarding the licensing of medicines, including veterinary medicines under section 4 of the Medicines Act 1968. Similarly, the agency will have an input to decisions concerning the licensing of pesticides and other related products under the Food and Environment Protection Act 1985. In both cases the agency has the ability to nominate a member to various government advisory committees on medicines, veterinary medicines and pesticides. Even most importantly in relation to the Environmental Protection Act 1990, which is concerned with preventing and minimising any damage to the environment which may result from the escape or release of genetically modified organisms, the new agency will be given a role in relation to regulations controlling the import, acquisition, release or marketing of any GMO and related matters. The Food Standards Agency will have the ability to act jointly with the Secretary of State for Health, in addition to the Secretary of State acting alone, in considering any exemptions for the risk assessment or notification requirements for GMOs and exempting any GMOs from the consent requirements relating to the same issues. The agency under Schedule 3 will be required to be consulted on any application or authorisation to release, market, import or require GMOs and on any regulations relating to fees and charges for applications under the Environmental Protection Act for deliberate release of GMOs.

Publication of advice

In establishing an independent food agency one of the main planks of the Blair government has been that such an agency will act in an open and transparent fashion. Section 19 of the 1999 Act therefore empowers the agency to publish advice given by it in accordance with its general functions of developing a food policy based on the protection of consumer health. Similarly, information obtained as a result of its various observations or enforcement monitoring should also be made public. The agency's ability to publish any of its advice to ministers will be an important factor in enhancing its influence and independence. Although ministers are not of course obliged to accept the agency's advice they would normally be expected to explain their reasons for not doing so. Obviously, such powers of transparency are subject to requirements of the Data Protection Act 1998, concerning personal information.

Before deciding to exercise its powers to publish its advice and information the agency must consider whether the "public interest" in the publication of the advice or information in question is outweighed by any considerations of confidentiality attaching to it. Section 19 should be read also in connection with section 25 of the 1999 Act which enables the Secretary of State for Health to make orders for the purpose of relaxing or over-riding any prohibitions on the disclosure of information contained in other food legislation that would otherwise prevent the agency from obtaining or publishing information in carrying out its functions under the 1999 Act. This section therefore provides a means for dealing with those situations where the agency would be unnecessarily limited in its ability to carry out its functions effectively if it was prohibited from disclosing information.

Another important power and function which is given to the Food Standards Agency is the ability to give advice, information and assistance to local authorities and other public bodies including health authorities on the management and control of outbreaks of food-borne illness (*e.g.* Salmonella, E.Coli 0157). Such guidance might, for example, include guidance in tracing the food-related source of any outbreak or in the speed in which action needs to be taken to limit the spread of food poisoning. These new powers which are provided to the agency under section 20 of the 1999 Act are very important functions especially in the aftermath of the BSE crisis and the dioxin contamination food scare in Belgium. Also, section 27 of the 1999 Act permits the Secretary of State for Health to make regulations to set up a notification scheme for the results of laboratory tests for food-borne organisms. This means that if a laboratory finds evidence that a person may have been exposed to certain pathogens that are capable of causing illness and are commonly transmitted through food, it will be required to report it to the central authorities. This information will improve data collection on types of food-borne disease. It will enable the Food Standards Agency to better understand patterns of the instance and prevalence of food-borne diseases. The pathogens initially expected to be covered by a notification scheme are Salmonella, E.Coli 0157 and Campylobacter.

Financing the Agency

The White Paper on a Food Standards Agency (January 1998) noted that the UK government intended to develop proposals for a fee-based registration or licensing scheme for food businesses which would cover costs associated with the agency. In the Food Standards Bill that was placed before Parliament in June 1999 there was provision in the Bill for a levy to be placed on all food businesses which would contribute to the costs of the running of the Food Standards Agency.

This levy clause no longer exists in the Food Standards Act 1999 and instead section 39 states that the agency will be funded from money provided by Parliament as well as monies from the Scottish Parliament, the National Assembly for Wales and the Northern Ireland Assembly. The devolved administrations will obviously fund the food body which will be aligned with the Food Standards Agency in an overall devolved administration, aimed at enhancing food safety thoughout the United Kingdom. There is also provision for the agency to obtain funding from the various services that it may provide to both local authorities and to the public in general. It is important to note that the Food Standards Agency will be funded directly by parliamentary vote, rather than by means of a grant paid by the minister. This is a very important stipulation as Parliament is less likely to make cut-backs in expenditure for food safety, compared to a Secretary of State or Minister who may be under pressure from the Treasury to cut back on various services within the ambit of his portfolio. Once appointments are made to the posts of chief executives and board members of the Food Standards Agency, the government's intention is to commence the work of the agency through statutory instrument in order that the Agency's functions and powers will come into operation as of 1 April 2000.

8.3 **Food Safety Authority of Ireland**

The Food Safety Authority of Ireland is a national independent agency, aiming to provide reassurance on food safety to consumers of food in Ireland and to consumers of Irish food abroad. In 1995, the Minister for Health established a Food Safety Advisory Board to advise his department on issues relating to food safety, nutrition, food law and scientific co-operation. Subsequently, in 1996 the government decided to establish the Food Safety Authority as an independent food agency to reassure consumers that the best hygiene and food safety standards/practices are observed in the Irish food industry. The Food Safety Advisory Board merged all its functions into the Food Safety Authority of Ireland on 1 January 1998. Subsequently, legislation was introduced in the form of the Food Safety Authority of Ireland Act, 1998. Under this Act the responsibilities of the Authority include:

- advice to consumers on food safety issues and food related diseases and risks;

- the inspection, approval, licensing and registration of premises involved in food production and distribution. This is at manufacturing, wholesale and retail levels;
- inspecting and sampling food;
- advising the National Disease Surveillance Unit and the Department of Health on potential food threats;
- undertaking food safety research, promotion, co-ordinating food control services and compiling food related statistics;
- working closely with other government agencies involved in food related areas;
- co-ordinating a certification process and ensuring "farm to plate" traceability of meat/food products;
- representing Ireland where appropriate on food safety issues, for example at the WHO etc.

Since its establishment the Food Safety Authority of Ireland has set up a food safety help line and a range of consumer literature on issues such as E.Coli 0157, genetically modified foods etc. It has been instrumental in moves to develop a food-borne diseases' surveillance system in Ireland establishing pilot projects in Dublin, Cork and Waterford based on systems used in the United Kingdom and the United States.

Penalties and enforcement

As of 1 July 1999, responsibility for the enforcement of all food safety legislation in Ireland was transferred to the Food Safety Authority of Ireland. Up to 2,000 staff and some 50 government agencies (local authorities and health boards) will continue to carry out their food safety activities under service contract arrangements with the Food Safety Authority of Ireland. Under section 50 of the 1998 Act authorised officers under the direction of the Food Safety Authority of Ireland have powers to enter premises in order to inspect hygiene and other such related conditions. The authorised officer is also given powers to obtain any information from persons concerned as well as to take samples if he so wishes.

Any person who fails to comply with a requirement made by an authorised officer commits an offence under section 50 of the 1998 Act and will be liable on summary conviction to a fine not exceeding £100,000 or to imprisonment for a term not exceeding three months or to both or on conviction on indictment to a fine not exceeding £10,000 or to imprisonment for a term not exceeding 12 months or to both.

Where a person continues to contravene section 50 of the Act he shall be guilty of an offence on every day on which the contravention continues and for each such offence be liable to a fine on summary conviction not exceeding £350 or on conviction on indictment to a fine not exceeding £15,000. On the other hand, a person who obstructs or interferes with an authorised officer in the exercise of his powers under the Act will be guilty of an offence and should be liable on summary conviction to a fine not exceeding £1,500 or to imprisonment for a term not exceeding three months or to both.

Improvement notices and closure orders

Under section 52 of the 1990 Act if an authorised officer is of the opinion that any activity taking place at a food premises has the possibility to pose a risk to public health, he may, following consultation with the chief executive or other such officer of the Food Safety Authority, serve on the proprietor of the food premises an "**improvement notice**". This notice shall identify the activity or defect in the premises giving rise to the risk and require that remedial action be taken and if appropriate shall specify the nature or details of such remedial action. The notice will also specify a time limit by which the action should be completed or implemented and may include other requirements as are necessary. The improvement notice when served on the proprietor of a food premises shall be effective immediately. If such an improvement notice is not complied with to the satisfaction of the authorised officer or the Food Safety Authority of Ireland, the authority may seek an "**improvement order**" from the district court directing the proprietor of the food premises to comply with the improvement notice. If an improvement order is not complied with within the time specified the authority may then serve a closure order.

Under section 52 of the 1998 Act there is provision for the authority to serve a "**closure order**" at any time in respect of any food premises which are subject to an improvement notice even if they have not sought an improvement order from the district court, once the authorised officer is of the opinion that the circumstances require the service of a closure order. The introduction of a closure order is specifically dealt with in section 53 of the 1998 Act. In general, the authorised officer should be of the opinion that there is likely to be a grave and immediate danger to public health if a particular food premises is not closed immediately. In the case of a closure order there is provision for the proprietor of a food premises to appeal the decision.

Under section 53(5)(a) a person who is aggrieved by a closure order may within the period of seven days beginning on the day on which the closure order is served on him appeal against the order to a judge of the district court in the district court district in which the order was served.

Where a closure order has been served and activities are carried on in contravention of the order, an application may be made to the High Court by the authorised officer prohibiting the continuance of the activities in the food premises concerned. In such a case the High Court when considering the matter may make such interim or interlocutory orders as it considers appropriate as well as introducing terms and conditions including payment of costs. Such an order was granted by the High Court on 15 December 1999 preventing a Co Limerick farmer continuing to use a premises at Gortnadromin, Pallasgreen, Co Limerick as an abattoir, since the Food Safety Authority of Ireland believed that the premises posed a grave and serious danger to public health. This is one of the first actions taken by the FSAI under the powers provided to it under the 1998 Act.[3]

Finally, the Food Safety Authority of Ireland has powers under section 54 of the 1990 Act to serve on a proprietor of a food business a **"prohibition order"** specifying that a particular assignment, class, batch or item of food should be withdrawn from sale where it may be the case that this matter involves a contravention of particular Irish food legislation. In such a situation, the prohibition order will ensure that the food is not used for human consumption, is recalled from sale or distribution or is detained or destroyed in a manner prescribed by the authorised officer. A person who is aggrieved by a prohibition order may, within the period of seven days beginning on the day on which the order is served on him, appeal against the order to a judge of the district court in the district court area where the prohibition order was served. The judge may confirm the prohibition order or cancel the prohibition as he sees fit. As is the case in a section 53 action there is also provision under section 54 where a prohibition order has been served and activities are carried on in contravention of the prohibition order, for the Food Safety Authority to make an application to the High Court to ensure that the prohibition order is enacted. The High Court in such a situation may make such interlocutory orders, as it considers appropriate and introduce terms and conditions including the issue of payment of costs.

---

3 *Irish Times*, 16 Dec 1999.

The Food Safety Authority of Ireland Act 1998 lists numerous pieces of EU legislation that the authority will have to ensure are implemented adequately in Ireland. The authority will also have the responsibility of co-operating with other national food agencies within the EU in the area of food safety research. There will also be a necessity for the Food Safety Authority to co-operate with a new North/South implementation body entitled the Food Safety Promotion Board that is being established in connection with the Belfast Agreement. This will also entail a degree of co-operation with the UK Food Standards Agency since the North/South body (Food Safety Promotion Board) comes under the aegis of the Food Standards Agency as specified in section 34 of the Food Standards Act 1999. Already, the Food Safety Authority of Ireland has been very involved in giving advice to the government, local health authorities and consumers on particular food scares. The Food Safety Authority of Ireland Act gives this new government agency strong powers to act as the "policemen" of the food industry in Ireland, in order to ensure that consumers can have confidence in the safety of the food they purchase. If the authority uses its powers wisely, it has the potential to play a major role in sustaining the viability of the Irish food industry, both at home and abroad.

8.4 **Conclusions**

By establishing a Food Standards Agency, the UK government is putting the interests of consumers first but it knows that the trust of food producers, manufacturers and retailers is required if it is to be successful in establishing a truly independent food agency. The remit given to the Food Standards Agency in the Food Standards Act 1999 is vast, but once the agency can prioritise its work it will have every chance of making a fundamental contribution to enhancing the importance of the existing body of UK food law. The government was right to discard the idea of funding the Food Standards Agency out of a levy to be paid by all sectors of the food industry. As the agency will now be funded almost completely from government sources, its independence will be maintained, which would not necessarily have been the case with the levy system.

The UK Food Standards Agency, as well as the Food Safety Authority of Ireland will have to contend with the certainty that the European Union will be establishing a European Food Authority by 2002. National food agencies may therefore find it difficult to balance their role with that of the European Food Authority. There may be possible "turf wars" between what the national agency perceives as being within its remit and what the European agency believes its role to be. This is likely to be more problematic for the UK agency as it will be trying to find its feet at the same time as the European Food Authority is being established. Even though there may be these teething problems the different agencies have one aim: to enhance food safety. In the long term, therefore, the UK Food Standards Agency and the European Food Authority are likely to cooperate fully together in order to protect the consumer health of both UK and European consumers.

Chapter 9

· **Conclusions** ·

Summary

- BSE crisis catalyst for change in UK and EU food law
- Weaknesses in the enforcement of food law in the UK
- Food Standards Agency – will enhance and close "loopholes" in enforcing food law
- Liability issue becoming ever more important in relation to food law
- EU White Paper on Food Safety (January 2000)
- Food law of growing importance to consumers, ministers and the food industry

9.1 **Introduction**

This book began with the Food Safety Act 1990 and ends with the Act establishing a Food Standards Agency, the Food Standards Act 1999. In the ensuing decade, there is little doubt that the importance of the existing body of UK food law has increased. In 1990, there were few UK lawyers specialising in food law; in general such lawyers would have dealt with consumer protection law and probably general commercial law. It is therefore ironic that the BSE crisis which began in the United Kingdom and rocked the foundation of the Conservative government as well as the European Union, has "spawned" a new legal specialisation – food law – both in the United Kingdom and throughout Europe. The BSE crisis has brought about radical changes in the entire EU regulatory regime for foodstuffs, which is now impinging on the United Kingdom.

The Blair government for all its belief in the importance of the consumers' "voice" has found that questions of food safety have the propensity to undermine the whole functioning of executive government. The Labour government has had to contend with a sustained debate on the safety of genetically modified foods in 1999 and at the end of the year its European policy regarding the ending of the EU ban on the export of British beef to Europe had been completely undermined by the French government. It must now wait for maybe two years for a judgment from the European Court of Justice in Luxembourg, before its policy on British beef can be vindicated. Food law is certainly proving to be of great importance in the UK body politic and it is likely to become ever more important for the food industry itself due to many of the liability issues discussed here.

9.2 **UK food law and its enforcement**

As has been mentioned previously, the Food Safety Act 1990, is the primary piece of food legislation in the United Kingdom. Whereas the majority of food law is now enacted following the adoption of proposals at EU level, the enactment and enforcement of these EU measures could not adequately take place without the powers and provisions contained in the 1990 Act. The 1990 Act introduces a number of different food law offences and establishes the system whereby UK food law is enforced. It is important to note that despite the fact that various EU laws have been implemented into UK law by statutory instruments which include offences and penalties provisions, the majority of food law prosecutions will be initiated under the auspices of the Food Safety Act 1990, rather than the statutory instrument. Obviously, the 1990 Act is more descriptive, includes very heavy fines and since many prosecutions begin in the magistrates' court, it is a more substantial statute to be discussing before lay magistrates.

Enforcement of food legislation in the United Kingdom is the responsibility of local authorities. Each local authority employs different officers to ensure the different UK food laws are enforced. There are two main types of officer – environmental health officers and trading standards officers. The environmental health officer will usually enforce those aspects of UK food law concerned with food hygiene and controls on food unfit for human consumption as outlined in the Food Safety Act 1990. The trading standards officer will be involved in the enforcement of those matters, as well as other controls relating to trade, labelling and composition of foodstuffs. Trading standards officers are also inspectors for weights and measures legislation. Veterinary supervision is also required in certain areas where animals enter the food chain, *e.g.* inspection at slaughterhouses.

The administration of food law varies throughout the United Kingdom, therefore, trading standards in England and Wales are allocated to county councils while environmental health controls are allocated to district councils. In metropolitan areas, both responsibilities are covered by metropolitan district councils and in London by borough councils. In Scotland district councils cover both responsibilities. With such a varied structure of local government, there has always been scope for a different level of enforcement around the country. To help achieve a uniform interpretation and application of law, the body LACTOS (Local Authorities Co-ordinating Body on Food and Trading Standards) assists to ensure that trading standards officers and environmental health officers apply the law in a uniform way. LACOTS has established a system whereby most companies only have to deal with one local authority ("the home authority") and any complaints are discussed with that authority. It is important, however, to remember that under the Food Safety Act 1990 and as demonstrated in the case law, in certain circumstances, trading standards officers and environmental officers can investigate matters outside their local authority jurisdiction.

Despite this elaborate structure for the enforcement of UK food law, the past decade has shown that there are gaps and loopholes in this system. The Pennington Report (1997) which examined the circumstances surrounding an outbreak of E.coli in Scotland resulting in 20 deaths and 500 cases of illness, amply points to the inadequacies of the present enforcement system. The report advocated increased use of HACCP systems throughout the entire food chain and recommended that in future the government in relation to food law should legislate "from the farm to the fork", in order to adequately protect consumer health. This report was instrumental in the background discussions for the government's White Paper on a Food Standards Agency (January 1998), which subsequently brought about the enactment in the statute books of the Food Standards Act 1999. This 1999 Act will strengthen the enforcement of UK food law in the coming years, since the Food Standards Agency has been given considerable powers to audit and monitor the performance of local enforcement authorities. It is hoped that this new oversight by the Food Standards Agency will locate and fill the gaps and loopholes in the existing system. The Agency will then be able to anticipate and prevent outbreaks of food poisoning occurring by strengthening the overall level of enforcement of UK food law both in local enforcement authorities and most importantly throughout the entire food production chain.

9.3 **Liability issues**

If there is one lesson that the food industry has learnt from the past decade of food scares such as the BSE crisis, it is that questions of liability are becoming ever more important to their business. As this book has demonstrated there are a number of existing laws on product liability, product safety and general food laws which all place obligations on food manufacturers or retailers to provide "safe" food to the consumer. If the food manufacturer or retailer fails in his duties he can be held liable under the specific liability laws or the Food Safety Act 1990.

As the BSE crisis undermined the entire EU regulatory regime for foodstuffs, there have been many calls for the introduction of new EU-wide liability legislation to assist in the protection of consumers' interests. The Product Liability Directive has been extended to include primary agricultural products and this will come into force in UK law as of December 2000. Before the ink is dry on that proposal the European Commission has published a Green Paper on Product Liability, which will begin a process where the entire "strict liability" regime established by that Directive may undergo radical change. If that were not enough, the European Commission is also reviewing the liability regime established by the General Product Safety Directive, so there are likely to be major changes there as well. Finally, there is a debate beginning in many Member States about the question of liability in the case of genetically modified plants/foods. The UK government has asked the EU to look into this issue and there is Private Member's Bill introduced by Alan Simpson MP (Labour) before the Westminster Parliament. Whatever may become of these initiatives, will be dependent on the recently introduced legal class actions taken in the United States, against large biotech companies like Monsanto.

There is little doubt that the coming year will produce many changes in UK food law, many coming from proposals contained in the Commission's White Paper on Food Safety (January 2000). The European Union has agreed to establish a European Food Authority which will also impinge on activities in the United Kingdom, especially the establishment of its own independent food agency in the guise of the Food Standards Agency. These matters will also impinge on food law issues in Ireland, where the food industry is of the utmost importance to the Irish economy and successive Irish governments have worked hard to ensure that food safety is taken seriously by all the players in the food industry – producers, manufacturers and government enforcement agencies – since if Irish food products get a bad name in terms of safety, they will be denied access to various markets, as the example of British beef shows. The British case also demonstrates that once a food product is "locked-out" of a market, it can be very difficult to get back into the same market in the future.

European food law is no longer concerned solely with the free movement of foodstuffs throughout the internal market, as was the case in the 1980s. The BSE crisis has meant that issues of food safety and consumer health have become paramount in relation to any new proposals for food legislation emerging from the European Commission. Similarly, UK food law which in the early 1990s was concerned with establishing an adequate food control system for food products in order that consumers could be certain about the safety of the food they purchased, it is now concentrating on enforcement and liability issues. Enforcement issues will be covered in much of the work of the new Food Standards Agency, so that the control system established by the Food Safety Act 1990, works better and increases the safety of food. In relation to liability issues, there is presently a body of UK law, but it is certain to increase in the foreseeable future. Those in the food industry who presently and in the future neglect questions of liability with regard to food, do so at their peril. They will not only be neglecting their consumers, but they will be putting their own businesses at risk. Food safety and product liability will therefore continue to be the important issues for the food industry in the foreseeable future.

Appendix 1

· Food Standards Act 1999 ·

1999 Chapter 28

An Act to establish the Food Standards Agency and make provision as to its functions; to amend the law relating to food safety and other interests of consumers in relation to food; to enable provision to be made in relation to the notification of tests for food-borne diseases; to enable provision to be made in relation to animal feedingstuffs; and for connected purposes.

[11th November 1999]

BE IT ENACTED by the Queen's most Excellent Majesty, by and with the advice and consent of the Lords Spiritual and Temporal, and Commons, in this present Parliament assembled, and by the authority of the same, as follows:-

The Food Standards Agency

The Food Standards Agency.

1.—(1) There shall be a body to be called the Food Standards Agency or, in Welsh, yr Asiantaeth Safonau Bwyd (referred to in this Act as "the Agency") for the purpose of carrying out the functions conferred on it by or under this Act.

(2) The main objective of the Agency in carrying out its functions is to protect public health from risks which may arise in connection with the consumption of food (including risks caused by the way in which it is produced or supplied) and otherwise to protect the interests of consumers in relation to food.

(3) The functions of the Agency are performed on behalf of the Crown.

Appointment of members etc.

2.—(1) The Agency shall consist of a chairman and deputy chairman and not less than eight or more than twelve other members, of whom—

    (a) one member shall be appointed by the National Assembly for Wales;

    (b) two members shall be appointed by the Scottish Ministers;

    (c) one member shall be appointed by the Department of Health and Social Services for Northern Ireland; and

    (d) the others shall be appointed by the Secretary of State.

(2) The chairman and deputy chairman shall be appointed by the appropriate authorities acting jointly and, before appointing a person as one of

the other members of the Agency the authority making the appointment shall consult the other appropriate authorities.

(3) Before appointing a person as chairman, deputy chairman or member of the Agency, the authorities or authority making the appointment shall—

> (a) have regard to the desirability of securing that a variety of skills and experience is available among the members of the Agency (including experience in matters related to food safety or other interests of consumers in relation to food); and
>
> (b) consider whether any person it is proposed to appoint has any financial or other interest which is likely to prejudice the exercise of his duties.

(4) Schedule 1 (constitution etc. of the Agency) has effect.

Appointment of chief executive and directors.

3.—(1) A chief executive shall be appointed for the Agency.

(2) The chief executive shall be responsible for (among other things) securing that the activities of the Agency are carried out efficiently and effectively.

(3) The first appointment under subsection (1) shall be made by the appropriate authorities acting jointly; and subsequent appointments shall be made by the Agency, subject to the approval of each of those authorities.

(4) Directors shall be appointed for Wales, for Scotland and for Northern Ireland, each of whom shall be responsible under the chief executive for (among other things) securing that the activities of the Agency in Wales, Scotland or Northern Ireland (as the case may be) are carried out efficiently and effectively.

(5) The first appointment under subsection (4) for Wales, for Scotland and for Northern Ireland shall be made by the appropriate authority for that part of the United Kingdom; and subsequent appointments shall be made by the Agency, subject to the approval of that authority.

(6) The chief executive and the directors appointed under subsection (4) shall hold and vacate office in accordance with the terms of their appointments.

Annual and other reports.

4.—(1) The Agency shall prepare a report on its activities and performance during each financial year.

(2) The Agency shall, as soon as possible after the end of each financial year, lay its report for that year before Parliament, the National Assembly for Wales, the Scottish Parliament and the Northern Ireland Assembly.

(3) The Agency may from time to time lay other reports before any of those bodies.

Advisory committees.

5.—(1) There shall be established an advisory committee for Wales, an advisory committee for Scotland and an advisory committee for Northern

Ireland for the purpose of giving advice or information to the Agency about matters connected with its functions (including in particular matters affecting or otherwise relating to Wales, Scotland or Northern Ireland, as the case may be).

(2) The Secretary of State may, after consulting the Agency, direct that an advisory committee for, or for any region of, England shall be established for the purpose of giving advice or information to the Agency about matters connected with its functions (including in particular matters affecting or otherwise relating to the area for which the committee is established).

(3) The Agency may, after consulting the appropriate authorities, establish other advisory committees for the purpose of giving advice or information to the Agency about matters connected with its functions.

(4) Schedule 2 (which contains supplementary provisions about advisory committees) has effect.

General functions in relation to food

Development of food policy and provision of advice, etc. to public authorities.

6.—(1) The Agency has the function of—
  - (a) developing policies (or assisting in the development by any public authority of policies) relating to matters connected with food safety or other interests of consumers in relation to food; and
  - (b) providing advice, information or assistance in respect of such matters to any public authority.

(2) A Minister of the Crown or government department, the National Assembly for Wales, the Scottish Ministers or a Northern Ireland Department may request the Agency to exercise its powers under this section in relation to any matter.

(3) It is the duty of the Agency, so far as is reasonably practicable, to comply with any such request.

Provision of advice, information and assistance to other persons.

7.—(1) The Agency has the function of—
  - (a) providing advice and information to the general public (or any section of the public) in respect of matters connected with food safety or other interests of consumers in relation to food;
  - (b) providing advice, information or assistance in respect of such matters to any person who is not a public authority.

(2) The function under subsection (1)(a) shall be carried out (without prejudice to any other relevant objectives) with a view to ensuring that members of the public are kept adequately informed about and advised in respect of matters which the Agency considers significantly affect their capacity to make informed decisions about food.

Acquisition and review of information

8.—(1) The Agency has the function of obtaining, compiling and keeping under review information about matters connected with food safety and other interests of consumers in relation to food.

(2) That function includes (among other things)—

(a) monitoring developments in science, technology and other fields of knowledge relating to the matters mentioned in subsection (1);

(b) carrying out, commissioning or co-ordinating research on those matters.

(3) That function shall (without prejudice to any other relevant objectives) be carried out with a view to ensuring that the Agency has sufficient information to enable it to take informed decisions and to carry out its other functions effectively.

General functions in relation to animal feedingstuffs

General functions in relation to animal feedingstuffs.

9.—(1) The Agency has the same general functions in relation to matters connected with the safety of animal feedingstuffs and other interests of users of animal feedingstuffs as it has under sections 6(1), 7(1) and 8 in relation to matters connected with food safety and other interests of consumers in relation to food.

(2) Section 6(2) and (3) apply in relation to the Agency's powers under this section corresponding to those under section 6(1).

(3) Section 7(2), in its application to the Agency's function under this section corresponding to that under section 7(1)(a), applies with the substitution, for the word "members of the public" and "food", of the words "users of animal feedingstuffs" and "animal feedingstuffs".

(4) In this section "safety of animal feedingstuffs" means the safety of animal feedingstuffs in relation to risks to animal health which may arise in connection with their consumption.

Observations with a view to acquiring information

Power to carry out observations.

10.—(1) The Agency may, for the purpose of carrying out its function under section 8 or its corresponding function under section 9, carry out observations (or arrange with other persons for observations to be carried out on its behalf) with a view to obtaining information about—

(a) any aspect of the production or supply of food or food sources; or

(b) any aspect of the production, supply or use of animal feedingstuffs.

(2) Without prejudice to the generality of subsection (1), the information that may be sought through such observations includes information about—

(a) food premises, food businesses or commercial operations being carried out with respect to food, food sources or contact materials;

(b) agricultural premises, agricultural businesses or agricultural activities;

(c) premises, businesses or operations involved in fish farming; or

(d) premises, businesses or operations involved in the production, supply or use of animal feedingstuffs.

(3) In this section—

"agricultural activity" has the same meaning as in the Agriculture Act 1947 or, in Northern Ireland, the Agriculture Act (Northern Ireland) 1949;

"agricultural business" has the same meaning as in section 1 of the Farm Land and Rural Development Act 1988 or, in Northern Ireland, Article 3 of the Farm Business (Northern Ireland) Order 1988;

"agricultural premises" means any premises used for the purposes of an agricultural business; and

"fish farming" means the breeding, rearing or keeping of fish or shellfish (which includes any kind of crustacean or mollusc).

Power of entry for persons carrying out observations.

11.—(1) The Agency may authorise any individual (whether a member of its staff or otherwise) to exercise the powers specified in subsection (4) for the purpose of carrying out any observations under section 10 specified in the authorisation.

(2) No authorisation under this section shall be issued except in pursuance of a decision taken by the Agency itself or by a committee, sub-committee or member of the Agency acting on behalf of the Agency.

(3) An authorisation under this section shall be in writing and may be given subject to any limitations or conditions specified in the authorisation (including conditions relating to hygiene precautions to be taken while exercising powers in pursuance of the authorisation).

(4) An authorised person may, if it appears to him necessary to do so for the purpose of carrying out the observations specified in his authorisation—

(a) enter any premises at any reasonable hour;

(b) take samples of any articles or substances found on any premises;

(c) take samples from any food source found on any premises;

(d) inspect and copy any records found on any premises which relate to a business which is the subject of the observations (and, if they are kept in computerised form, require them to be made available in a legible form);

(e) require any person carrying on such a business to provide him with such facilities, such records or information and such other assistance as he may reasonably request;

but in this subsection "premises" does not include a private dwelling-house.

(5) An authorised person shall on request—

(a) produce his authorisation before exercising any powers under subsection (4); and

(b) provide a document identifying any sample taken, or documents copied, under those powers.

(6) The references in subsection (4)(d) and (e) to records include any records which—

(a) relate to the health of any person who is or has been employed in the business concerned; and

(b) were created for the purpose of assessing, or are kept for the purpose of recording, matters affecting his suitability for working in the production or supply of food or food sources (including any risks to public health which may arise if he comes into contact with any food or food source).

(7) If an authorised person who enters any premises by virtue of this section discloses to any person any information obtained on the premises with regard to any trade secret he is, unless the disclosure is made in the performance of his duty, guilty of an offence and liable on summary conviction to a fine not exceeding level 5 on the standard scale.

(8) A person who—

(a) intentionally obstructs a person exercising powers under subsection (4)(a), (b), (c) or (d);

(b) fails without reasonable excuse to comply with any requirement imposed under subsection (4)(e); or

(c) in purported compliance with such a requirement furnishes information which he knows to be false or misleading in any material particular or recklessly furnishes information which is false or misleading in any material particular;

is guilty of an offence and liable on summary conviction to a fine not exceeding level 5 on the standard scale.

(9) In this section "authorised person" means a person authorised under this section.

Monitoring of enforcement action

Monitoring of enforcement action.

12.—(1) The Agency has the function of monitoring the performance of enforcement authorities in enforcing relevant legislation.

(2) That function includes, in particular, setting standards of performance (whether for enforcement authorities generally or for particular authorities) in relation to the enforcement of any relevant legislation.

(3) Each annual report of the Agency shall contain a report on its activities during the year in enforcing any relevant legislation for which it is the enforcement authority and its performance in respect of—

(a) any standards under subsection (2) that apply to those activities; and

(b) any objectives relating to those activities that are specified in the statement of objectives and practices under section 22.

(4) The Agency may make a report to any other enforcement authority on their performance in enforcing any relevant legislation; and such a report may include guidance as to action which the Agency considers would improve that performance.

(5) The Agency may direct an authority to which such a report has been made—

(a) to arrange for the publication in such manner as may be specified in the direction of, or of specified information relating to, the report; and

(b) within such period as may be so specified to notify the Agency of what action they have taken or propose to take in response to the report.

Power to request information relating to enforcement action.

13.—(1) For the purpose of carrying out its function under section 12 in relation to any enforcement authority the Agency may require a person mentioned in subsection (2)—

(a) to provide the Agency with any information which it has reasonable cause to believe that person is able to give, or

(b) to make available to the Agency for inspection any records which it has reasonable cause to believe are held by that person or otherwise within his control (and, if they are kept in computerised form, to make them available in a legible form).

(2) A requirement under subsection (1) may be imposed on—

(a) the enforcement authority or any member, officer or employee of the authority, or

(b) a person subject to any duty under relevant legislation (being a duty enforceable by an enforcement authority) or any officer or employee of such a person.

(3) The Agency may copy any records made available to it in pursuance of a requirement under subsection (1)(b).

Power of entry for persons monitoring enforcement action.

14.—(1) The Agency may authorise any individual (whether a member of its staff or otherwise) to exercise the powers specified in subsection (4) for the purpose of carrying out its function under section 12 in relation to any enforcement authority.

(2) No authorisation under this section shall be issued except in pursuance of a decision taken by the Agency itself or by a committee, sub-committee or member of the Agency acting on behalf of the Agency.

(3) An authorisation under this section shall be in writing and may be given subject to any limitations or conditions specified in the authorisation (including conditions relating to hygiene precautions to be taken while exercising powers in pursuance of the authorisation).

(4) An authorised person may—

    (a) enter any premises mentioned in subsection (5) at any reasonable hour in order to inspect the premises or anything which may be found on them;

    (b) take samples of any articles or substances found on such premises;

    (c) inspect and copy any records found on such premises (and, if they are kept in computerised form, require them to be made available in a legible form);

    (d) require any person present on such premises to provide him with such facilities, such records or information and such other assistance as he may reasonably request.

(5) The premises which may be entered by an authorised person are—

    (a) any premises occupied by the enforcement authority;

    (b) any laboratory or similar premises at which work related to the enforcement of any relevant legislation has been carried out for the enforcement authority; and

    (c) any other premises (not being a private dwelling-house) which the authorised person has reasonable cause to believe are premises in respect of which the enforcement powers of the enforcement authority are or have been exercisable.

(6) The power to enter premises conferred on an authorised person includes power to take with him any other person he may consider appropriate.

(7) An authorised person shall on request—

    (a) produce his authorisation before exercising any powers under subsection (4); and

    (b) provide a document identifying any sample taken, or documents copied, under those powers.

(8) If a person who enters any premises by virtue of this section discloses to any person any information obtained on the premises with regard to any trade secret he is, unless the disclosure is made in the performance of his duty, guilty of an offence and liable on summary conviction to a fine not exceeding level 5 on the standard scale.

(9) Where —

    (a) the enforcement authority in relation to any provisions of the Food Safety Act 1990 (in this Act referred to as "the 1990 Act") or orders or regulations made under it is (by virtue of section 6(3) or (4) of that Act) a Minister of the Crown, the National Assembly for Wales, the Scottish Ministers or the Agency, or

    (b) the enforcement authority in relation to any provisions of the Food Safety (Northern Ireland) Order 1991 (in this Act referred to as "the 1991 Order") or orders or regulations made under it is (by virtue of Article 26(1A), (1B), (2), (3) or (3A) of that Order) a Northern Ireland Department or the Agency,

this section applies to that authority (in relation to its performance in enforcing those provisions) with the omission of subsection (5)(a).

(10) In this section "authorised person" means a person authorised under this section.

Meaning of "enforcement authority" and related expressions.

15.—(1) In sections 12 to 14 "relevant legislation" means—

(a) the provisions of the 1990 Act and regulations or orders made under it;

(b) the provisions of the 1991 Order and regulations or orders made under it; and

(c) the provisions of Part IV of the Agriculture Act 1970 and regulations made under that Part of that Act, so far as relating to matters connected with animal feedingstuffs.

(2) In those sections "enforcement authority" means—

(a) in the case of provisions of the 1990 Act or regulations or orders made under it, the authority by whom they are to be enforced (including a Minister of the Crown, the National Assembly for Wales, the Scottish Ministers or the Agency itself if, by virtue of section 6(3) or (4) of the 1990 Act, that authority is the enforcement authority in relation to those provisions);

(b) in the case of provisions of the 1991 Order and regulations or orders made under it, the authority by whom they are to be enforced (including a Northern Ireland Department or the Agency itself if, by virtue of the Order, it is the enforcement authority in relation to those provisions); and

(c) in the case of provisions of Part IV of the Agriculture Act 1970 (or regulations made under it), an authority mentioned in section 67 of that Act;

and "enforcement", in relation to relevant legislation, includes the execution of any provisions of that legislation.

(3) Any reference in those sections (however expressed) to the performance of an enforcement authority in enforcing any relevant legislation includes a reference to the capacity of that authority to enforce it.

Offences relating to sections sections 13 and 14.

16. —(1) A person who—

(a) intentionally obstructs a person exercising powers under section 14(4)(a), (b) or (c);

(b) fails without reasonable excuse to comply with any requirement imposed under section 13(1) or section 14(4)(d); or

(c) in purported compliance with such a requirement furnishes information which he knows to be false or misleading in any material particular or recklessly furnishes information which is false or misleading in any material particular;

is guilty of an offence.

(2) A person guilty of an offence under this section is liable on summary conviction to a fine not exceeding level 5 on the standard scale.

Other functions of the Agency

Delegation of powers to make emergency orders.

17.—(1) Arrangements may be made between the Secretary of State and the Agency authorising the Agency to exercise on behalf of the Secretary of State the power to make orders under—

    (a) section 1(1) of the Food and Environment Protection Act 1985 (emergency orders); and

    (b) section 13(1) of the 1990 Act (emergency control orders).

(2) The authority given by any such arrangements is subject to any limitations and conditions provided for in the arrangements.

(3) Where by virtue of any such arrangements the Agency is authorised to exercise a power, anything done or omitted to be done by the Agency in the exercise or purported exercise of the power shall be treated as done or omitted by the Secretary of State.

(4) Nothing in any such arrangements prevents the Secretary of State exercising any power.

(5) This section applies with the necessary modifications—

    (a) to any power mentioned in subsection (1) so far as it is exercisable by the National Assembly for Wales or the Scottish Ministers, and

    (b) to the power of a Northern Ireland Department to make orders under section 1(1) of the Food and Environment Protection Act 1985 or Article 12(1) of the 1991 Order,

as it applies to a power exercisable by the Secretary of State.

Functions under other enactments.

18.—(1) Schedule 3 (which contains provisions conferring functions under certain enactments on the Agency) has effect.

(2) Any amendment made by Schedule 3 which extends to Scotland is to be taken as a pre-commencement enactment for the purposes of the Scotland Act 1998.

Publication etc. by the Agency of advice and information.

19.—(1) The Agency may, subject to the following provisions of this section, publish in such manner as it thinks fit—

    (a) any advice given under section 6, 7 or 9 (including advice given in pursuance of a request under section 6(2));

    b) any information obtained through observations under section 10 or monitoring under section 12; and

    (c) any other information in its possession (whatever its source).

(2) The exercise of that power is subject to the requirements of the Data Protection Act 1998.

(3) That power may not be exercised if the publication by the Agency of the advice or information in question—

(a) is prohibited by an enactment;

(b) is incompatible with any Community obligation; or

(c) would constitute or be punishable as a contempt of court.

(4) Before deciding to exercise that power, the Agency must consider whether the public interest in the publication of the advice or information in question is outweighed by any considerations of confidentiality attaching to it.

(5) Where the advice or information relates to the performance of enforcement authorities, or particular enforcement authorities, in enforcing relevant legislation, subsection (4) applies only so far as the advice or information relates to a person other than—

(a) an enforcement authority, or

(b) a member, officer or employee of an enforcement authority acting in his capacity as such.

(6) Expressions used in subsection (5) and defined in section 15 have the same meaning as in that section.

(7) Except as mentioned above, the power under subsection (1) is exercisable free from any prohibition on publication that would apply apart from this section.

(8) In this section "enactment" means an enactment contained in, or in subordinate legislation made under, any Act, Act of the Scottish Parliament or Northern Ireland legislation.

(9) The Agency may also disclose to another public authority any advice or information mentioned in subsection (1); and the other provisions of this section apply in relation to disclosure under this subsection as they apply in relation to publication under that subsection.

Power to issue guidance on control of food-borne diseases.

20.—(1) The Agency may issue general guidance to local authorities or other public authorities on matters connected with the management of outbreaks or suspected outbreaks of food-borne disease.

(2) Guidance issued under this section must identify the authority or authorities to which it is addressed.

(3) The Agency shall publish any guidance issued under this section in such manner as it thinks fit.

(4) Any authority to whom guidance under this section is issued shall have regard to the guidance in carrying out any functions to which the guidance relates.

(5) In this section "food-borne disease" means a disease of humans which is capable of being caused by the consumption of infected or otherwise contaminated food.

(6) This section has effect without prejudice to any other powers of the Agency.

Supplementary powers.

21.—(1) The Agency has power to do anything which is calculated to facilitate, or is conducive or incidental to, the exercise of its functions.

(2) Without prejudice to the generality of subsection (1), that power includes power—

    (a) to carry on educational or training activities;

    (b) to give financial or other support to activities carried on by others;

    (c) to acquire or dispose of any property or rights;

    (d) to institute criminal proceedings in England and Wales and in Northern Ireland.

(3) The Agency may make charges for facilities or services provided by it at the request of any person.

General provisions relating to the functions of the Agency

Statement of general objectives and practices.

22.—(1) The Agency shall prepare and publish a statement of general objectives it intends to pursue, and general practices it intends to adopt, in carrying out its functions.

(2) The statement shall include the following among the Agency's general objectives, namely—

    (a) securing that its activities are the subject of consultation with, or with representatives of, those affected and, where appropriate, with members of the public;

    (b) promoting links with any of the following authorities with responsibilities affecting food safety or other interests of consumers in relation to food, namely—

        (i) government departments, local authorities and other public authorities;

        (ii) the National Assembly for Wales (and its staff) and Assembly Secretaries, the Scottish Administration and Northern Ireland Departments;

        with a view to securing that the Agency is consulted informally from time to time about the general manner in which any such responsibilities are discharged;

    (c) securing that records of its decisions, and the information on which they are based, are kept and made available with a view to enabling members of the public to make informed judgments about the way in which it is carrying out its functions,

and any other objectives (which may include more specific objectives relating to anything mentioned in paragraphs (a) to (c)) which are notified to the Agency by the appropriate authorities acting jointly.

(3) Nothing in subsection (2) prevents the inclusion in the statement of more specific objectives relating to anything mentioned in that subsection.

(4) The statement shall be submitted in draft to the appropriate authorities for their approval before it is published.

(5) The appropriate authorities acting jointly may approve the draft statement submitted to them with or without modifications (but they must consult the Agency before making any modifications).

(6) As soon as practicable after a statement is approved under subsection (5), the Agency shall—

    (a) lay a copy of the statement as so approved before Parliament, the National Assembly for Wales, the Scottish Parliament and the Northern Ireland Assembly; and

    (b) publish that statement in such manner as the appropriate authorities acting jointly may approve.

(7) The first statement under this section shall be submitted to the appropriate authorities within the period of three months beginning with the date of the first meeting of the Agency.

(8) The Agency may revise its current statement under this section; and subsections (2) to (6) apply to a revised statement as they apply to the first statement.

Consideration of objectives, risks, costs and benefits, etc.

23.—(1) In carrying out its functions the Agency shall pay due regard to the statement of objectives and practices under section 22.

(2) The Agency, in considering whether or not to exercise any power, or the manner in which to exercise any power, shall take into account (among other things)—

    (a) the nature and magnitude of any risks to public health, or other risks, which are relevant to the decision (including any uncertainty as to the adequacy or reliability of the available information);

    (b) the likely costs and benefits of the exercise or non-exercise of the power or its exercise in any manner which the Agency is considering; and

    (c) any relevant advice or information given to it by an advisory committee (whether or not given at the Agency's request).

(3) The duty under subsection (2)—

    (a) does not apply to the extent that it is unreasonable or impracticable for it to do so in view of the nature or purpose of the power or in the circumstances of the particular case; and

    (b) does not affect the obligation of the Agency to discharge any other duties imposed on it.

Directions relating to breach of duty or to international obligations.

24.—(1) If it appears to the Secretary of State that there has been a serious failure by the Agency—

    (a) to comply with section 23(1) or (2), or

    (b) to perform any other duty which he considers should have been performed by it,

he may give the Agency such directions as he may consider appropriate for remedying that failure.

    (2) The power under subsection (1) may also be exercised—

    (a) so far as it is exercisable in relation to Wales, by the National Assembly for Wales;

    (b) by the Scottish Ministers (in so far as it is exercisable by them within devolved competence or by virtue of an Order in Council made under section 63 of the Scotland Act 1998); and

    (c) so far as it is exercisable in relation to Northern Ireland, by the Department of Health and Social Services for Northern Ireland.

    (3) Directions under subsection (1) must include a statement summarising the reasons for giving them.

    (4) The Secretary of State may give the Agency such directions as he considers appropriate for the implementation of—

    (a) any obligations of the United Kingdom under the Community Treaties, or

    (b) any international agreement to which the United Kingdom is a party.

    (5) The power under subsection (4) may also be exercised—

    (a) by the National Assembly for Wales (in relation to implementation for which it is responsible);

    (b) by the Scottish Ministers (in relation to implementation within devolved competence or for which they have responsibility by virtue of an Order in Council under section 63 of the Scotland Act 1998); and

    (c) by the Department of Health and Social Services for Northern Ireland (in relation to implementation for which a Northern Ireland Department is responsible).

    (6) An authority proposing to give directions under this section shall consult the Agency and the other appropriate authorities before doing so.

    (7) If the Agency fails to comply with any directions under this section, the authority giving the directions may give effect to them (and for that purpose may exercise any power of the Agency).

    (8) If the Agency fails to comply with directions under subsection (1), the Secretary of State may, with the agreement of the other appropriate authorities, remove all the members of the Agency from office (and, until new appointments are made, may carry out the Agency's functions himself or appoint any other person or persons to do so).

    (9) Any directions given under this section shall be published in such manner as the authority giving them considers appropriate for the purpose of bringing the matters to which they relate to the attention of persons likely to be affected by them.

    (10) In this section "devolved competence" has the same meaning as in the Scotland Act 1998.

Power to modify enactments about disclosure of information.

25.—(1) If it appears to the Secretary of State that an enactment prohibits the disclosure of any information and is capable of having either of the effects mentioned in subsection (5) he may by order make provision for the purpose of removing or relaxing the prohibition so far as it is capable of having that effect.

(2) If it appears to the Scottish Ministers that an enactment prohibits the disclosure of any information and is capable of having either of the effects mentioned in subsection (5) the Scottish Ministers may by order make provision for the purpose of removing or relaxing the prohibition so far as it is capable of having that effect.

(3) The power under subsection (2) may not be exercised to make provision which would not be within the legislative competence of the Scottish Parliament.

(4) If it appears to the First Minister and deputy First Minister acting jointly that any enactment dealing with transferred matters (within the meaning of section 4(1) of the Northern Ireland Act 1998) prohibits the disclosure of any information and is capable of having either of the effects mentioned in subsection (5) they may by order make provision for the purpose of removing or relaxing the prohibition so far as it is capable of having that effect.

(5) The effects mentioned in subsections (1), (2) and (4) are that the enactment in question—

    (a) prevents the disclosure to the Agency of information that would facilitate the carrying out of the Agency's functions; or

    (b) prevents the publication by the Agency of information in circumstances where the power under section 19 would otherwise be exercisable.

(6) An order under this section may—

    (a) make provision as to circumstances in which information which is subject to the prohibition in question may, or may not, be disclosed to the Agency or, as the case may be, published by the Agency; and

    (b) if it makes provision enabling the disclosure of information to the Agency, make provision restricting the purposes for which such information may be used (including restrictions on the subsequent disclosure of the information by the Agency).

(7) This section applies in relation to a rule of law as it applies in relation to an enactment, but with the omission of—

    (a) subsection (5)(b) and any reference to the effect mentioned in subsection (5)(b); and

    (b) in subsection (6)(a), the words from "or, as" to the end.

(8) In this section "enactment" means an enactment contained in any Act (other than this Act) or Northern Ireland legislation passed or made before or in the same Session as this Act.

Miscellaneous provisions

Statutory functions ceasing to be exercisable by Minister of Agriculture, Fisheries and Food and Department of Agriculture for Northern Ireland.

26.—(1) The functions of the Minister of of Agriculture, Fisheries and Food under—

(a) Part I of the Food and Environment Protection Act 1985;

(b) the 1990 Act; and

(c) the Radioactive Substances Act 1993,

shall cease to be exercisable by that Minister.

(2) The functions of the Department of Agriculture for Northern Ireland under—

(a) Part I of the Food and Environment Protection Act 1985; and

(b) Part II of the 1991 Order (except Articles 8(7), 10(5) to (7), 11(5) to (10), 18(1), 22 and 25(2)(e) and Schedule 1),

shall cease to be exercisable by that Department.

(3) Subsections (1) and (2) do not affect enforcement functions under directions or subordinate legislation under the enactments mentioned in those subsections (or any power under those enactments to confer such functions in directions or subordinate legislation).

Notification of tests for food-borne disease.

27.—(1) Regulations may make provision for requiring the notification of information about tests on samples taken from individuals (whether living or dead) for the presence of—

(a) organisms of a description specified in the regulations; or

(b) any substances produced by or in response to the presence of organisms of a description so specified.

(2) A description of organisms may be specified in the regulations only if it appears to the authority making the regulations that those organisms or any substances produced by them—

(a) are capable of causing disease in humans; and

(b) are commonly transmitted to humans through the consumption of food.

(3) The power to make the regulations is exercisable for the purpose of facilitating the carrying out of functions of the Agency or any other public authority which relate to the protection of public health.

(4) The regulations shall, as respects each specified description of organisms—

(a) specify the information to be notified about them and the form and manner in which it is to be notified;

b) make provision for identifying the person by whom that information is to be notified; and

(c) specify the person to whom that information is to be notified;

but the regulations may not require a person to notify information which is not in his possession, or otherwise available to him, by virtue of his position.

(5) The regulations may—

(a) make provision as to the tests about which information is to be notified;

(b) require or permit the person specified under subsection (4)(c) to disclose any information to any other person or to publish it;

(c) restrict the purposes for which any information may be used (whether by the person so specified or by any other person);

(d) make provision with a view to ensuring that patient confidentiality is preserved;

(e) create exceptions from any provision of the regulations;

(f) create summary offences, subject to the limitation that no such offence shall be punishable with imprisonment or a fine exceeding level 5 on the standard scale.

(6) Before making regulations under this section the authority making them shall consult the Agency and such organisations as appear to the authority to be representative of interests likely to be substantially affected by the regulations.

(7) Any consultation undertaken before the commencement of subsection (6) shall be as effective, for the purposes of that subsection, as if undertaken after that commencement.

(8) The power to make regulations under this section is exercisable—

(a) as respects tests carried out in England, by the Secretary of State;

(b) as respects tests carried out in Wales, by the National Assembly for Wales;

(c) as respects tests carried out in Scotland, by the Scottish Ministers; and

(d) as respects tests carried out in Northern Ireland, by the Department of Health and Social Services for Northern Ireland.

Arrangements for sharing information about food-borne zoonoses.

28.—(1) The Agency and each authority to which this section applies with responsibility for any matter connected with food-borne zoonoses shall make arrangements with a view to securing (so far as reasonably practicable) that any information relating to food-borne zoonoses in the possession of either of them is furnished or made available to the other.

(2) The authorities to which this section applies are Ministers of the Crown, the National Assembly for Wales, Scottish Ministers and Northern Ireland Departments.

(3) Arrangements under this section may also include arrangements for co-ordinating the activities of the Agency and the authority concerned in relation to matters connected with food-borne zoonoses.

(4) Arrangements under this section shall be kept under review by the Agency and the authority concerned.

(5) In this section "food-borne zoonosis" means any disease of, or organism carried by, animals which constitutes a risk to the health of humans through the consumption of, or contact with, food.

Consultation on veterinary products.

29.—(1) The Minister of Agriculture, Fisheries and Food, and each Secretary of State having responsibility for any matters connected with the regulation of veterinary products, shall consult the Agency from time to time about the general policy he proposes to pursue in carrying out his functions in relation to those matters.

(2) In this section "veterinary products" means—

    (a) veterinary drugs, as defined in section 132(1) of the Medicines Act 1968;

    (b) veterinary medicinal products, as defined in Article 1(2) of Council Directive 81/851/EEC (including products manufactured from homeopathic stock);

    (c) medicated feedingstuffs, as defined in Article 1(2) of Council Directive 81/851/EEC;

    (d) zootechnical products, as defined in regulation 2(1) of the Feedingstuffs (Zootechnical Products) Regulations 1999.

(3) The Minister or the Secretary of State concerned may disclose any information to the Agency (including information obtained by or furnished to him in pursuance of any enactment) relating to matters connected with the regulation of veterinary products.

(4) This section applies to the Department of Health and Social Services for Northern Ireland and the Department of Agriculture for Northern Ireland as it applies to the Minister of Agriculture, Fisheries and Food.

Animal feedingstuffs: Great Britain.

30.—(1) The Ministers may, for the purpose of regulating any animal feedingstuff or anything done to or in relation to, or with a view to the production of, any animal feedingstuff, make an order under this section.

(2) An order under this section is one which applies, or makes provision corresponding to, any provisions of the 1990 Act (including any power to make subordinate legislation or to give directions), with or without modifications.

(3) Such an order may be made by reference to the 1990 Act as it stands immediately before this Act is passed or as it stands following any amendment or repeal made by this Act.

(4) Such an order under this section may make provision with a view to protecting animal health, protecting human health or for any other purpose which appears to the Ministers to be appropriate.

(5) The provision which may be made in an order under this section by virtue of section 37(1)(a) includes provision amending or repealing any enactment or subordinate legislation.

(6) Before making such an order, the Ministers shall—
  (a) consult such organisations as appear to them to be representative of interests likely to be substantially affected by the order; and
  (b) have regard to any advice given by the Agency.

(7) Any consultation undertaken before the commencement of subsection (6) shall be as effective, for the purposes of that subsection, as if undertaken after that commencement; and any consultation undertaken by the Agency may be treated by the Ministers as being as effective for those purposes as if it had been undertaken by them.

(8) In this section "the Ministers" means—
  (a) in the case of an order extending to England and Wales, the Secretary of State and the Minister of Agriculture, Fisheries and Food, acting jointly;
  (b) in the case of an order extending to Scotland, the Scottish Ministers.

Animal feedingstuffs: Northern Ireland.

31.—(1) The Department of Agriculture for Northern Ireland and the Department of Health and Social Services for Northern Ireland acting jointly shall have the same power to make provision by order for Northern Ireland by reference to the 1991 Order as the Ministers have by virtue of section 30 to make provision by order for England and Wales or Scotland by reference to the 1990 Act.

(2) Subsections (6) and (7) of section 30 apply in relation to an order under this section as they apply to an order under that section.

Modification of certain provisions of this Act.

32.—(1) Her Majesty may by Order in Council make such provision as She considers appropriate for modifying—
  (a) the functions exercisable under this Act by any of the appropriate authorities (including functions exercisable jointly by two or more of them);
  (b) the powers under this Act of either House of Parliament, the Scottish Parliament or the Northern Ireland Assembly; or
  (c) the constitution of the Agency.

(2) Without prejudice to the generality of subsection (1), provision made under paragraph (a) or (b) of that subsection may—
  (a) confer on any one or more of the appropriate authorities functions (including powers to make subordinate legislation) which relate to anything connected with the Agency or its activities;
  (b) confer powers on either House of Parliament, the Scottish Parliament or the Northern Ireland Assembly.

(3) Where provision is made under subsection (1)(a) or (b), the provision which may be made in the Order by virtue of section 37(1)(a) includes provision modifying functions of, or conferring functions on, the Agency or any other person in connection with any one or more of the appropriate authorities or with any body mentioned in subsection (1)(b).

(4) For the purposes of subsection (1)(c) the reference to the constitution of the Agency is a reference to the subject-matter of sections 2 to 5 and 39(7) (together with Schedules 1, 2 and 4).

(5) The provision which may be made by an Order under this section does not include provision modifying this section or section 33 (except that where provision is made under subsection (1)(c) the Order may make consequential amendments to subsection (4)).

(6) No recommendation shall be made to Her Majesty in Council to make an Order under this section unless the Agency has been consulted.

Consequences of Agency losing certain functions.
    33.—(1) This section applies if—
    (a) the Scottish Parliament passes an Act providing for any functions of the Agency to be no longer exercisable in or as regards Scotland; or
    (b) the Northern Ireland Assembly passes an Act providing for any functions of the Agency to be no longer exercisable in or as regards Northern Ireland.
  (2) Her Majesty may by Order in Council make provision—
    (a) modifying this or any other Act as She considers necessary or expedient in consequence of the functions concerned being no longer exercisable by the Agency in or as regards Scotland or Northern Ireland;
    (b) for the transfer of any property, rights and interests of the Agency falling within subsection (3);
    (c) for any person to have such rights or interests in relation to any property, rights or interests falling within subsection (3) as She considers appropriate (whether in connection with a transfer or otherwise); or
    (d) for the transfer of any liabilities of the Agency falling within subsection (4).
  (3) Property, rights and interests fall within this subsection if they belong to the Agency and appear to Her Majesty—
    (a) to be held or used wholly or partly for or in connection with the exercise of any of the functions concerned, or
    (b) not to be within paragraph (a) but, when last held or used for or in connection with the exercise of any function, to have been so held or used for or in connection with the exercise of any of the functions concerned.
  (4) Liabilities of the Agency fall within this subsection if they appear to Her Majesty to have been incurred wholly or partly for or in connection with the exercise of any of the functions concerned.
  (5) An Order under this section may make provision for the delegation of powers to determine anything required to be determined for the purposes of provision made under subsection (2)(b), (c) or (d).

(6) No recommendation shall be made to Her Majesty in Council to make an Order under this section unless the Agency has been consulted.

Duty to take account of functions of the Food Safety Promotion Board.

34.—(1) The Agency must—

    (a) take account of the activities of the Food Safety Promotion Board in determining what action to take for the purpose of carrying out its functions; and

    (b) consult that Board from time to time with a view to ensuring so far as is practicable that the activities of the Agency do not unnecessarily duplicate the activities of the Board.

(2) Nothing in this Act affects the functions of the Food Safety Promotion Board.

Devolution in Scotland and Northern Ireland.

35.—(1) For the purposes of—

    (a) section 23(2)(b) of the Scotland Act 1998 (power of Scottish Parliament to require persons outside Scotland to attend to give evidence or produce documents); and

    (b) section 70(6) of that Act (accounts prepared by cross-border bodies),

the Agency shall be treated as a cross-border public authority (within the meaning of that Act).

(2) It is not outside the legislative competence of the Scottish Parliament, by virtue of the reservation of matters relating to the constitution mentioned in paragraph 1 of Schedule 5 to that Act, to remove, alter or confer relevant functions of the Agency which are exercisable in or as regards Scotland.

(3) Nothing in subsection (2) affects any legislative competence of the Scottish Parliament apart from this section.

(4) Relevant functions of the Agency in relation to Northern Ireland shall be regarded as functions of a Minister of the Crown for the purposes of paragraph 1(a) of Schedule 2 to the Northern Ireland Act 1998 (excepted matters).

(5) In this section "relevant functions of the Agency" means functions relating to, or to matters connected with—

    (a) food safety or other interests of consumers in relation to food; or

    (b) the safety of animal feedingstuffs or other interests of users of animal feedingstuffs.

Final provisions

Interpretation.

36.—(1) In this Act—

    "Agency" means the Food Standards Agency;

"animal feedingstuff" means feedingstuff for any description of animals, including any nutritional supplement or other similar substance which is not administered through oral feeding;

"appropriate authorities" means the Secretary of State, the National Assembly for Wales, the Scottish Ministers and the Department of Health and Social Services for Northern Ireland;

"Food Safety Promotion Board" means the body of that name established by the agreement establishing implementation bodies done at Dublin on 8th March 1999 between the Government of the United Kingdom and the Government of Ireland;

"the 1990 Act" means the Food Safety Act 1990; and

"the 1991 Order" means the Food Safety (Northern Ireland) Order 1991.

(2) Any reference in this Act to "the appropriate authority", in relation to Wales, Scotland or Northern Ireland, is a reference to the National Assembly for Wales, the Scottish Ministers or the Department of Health and Social Services for Northern Ireland (as the case may be).

(3) In this Act the expression "interests of consumers in relation to food" includes (without prejudice to the generality of that expression) interests in relation to the labelling, marking, presenting or advertising of food, and the descriptions which may be applied to food.

(4) Expressions used-

(a) as regards England and Wales and Scotland, in this Act and in the 1990 Act, or

(b) as regards Northern Ireland, in this Act and the 1991 Order,

have, unless the context otherwise requires, the same meaning in this Act as in that Act or that Order (except that in this Act "animal" includes any bird or fish).

(5) The purposes which may be specified in an order under section 1(3) of the 1990 Act (meaning of the term "premises" to include, for specified purposes, ships or aircraft of a description specified by order), or under the corresponding provision of Article 2(2) of the 1991 Order, include purposes relating to provisions of this Act.

Subordinate legislation.

37.—(1) Subordinate legislation under section 25, 27, 30, 31, 32 and 33—

(a) may contain such supplementary, incidental, consequential, transitional or saving provision as the person making it considers necessary or expedient;

(b) may make different provision for different purposes.

(2) Any power under this Act to make an order or regulations is exercisable—

(a) in the case of an order or regulations made by the First Minister and deputy First Minister or a Northern Ireland Department, by statutory rule for the purposes of the Statutory Rules (Northern Ireland) Order 1979; and

(b) in any other case, by statutory instrument.

(3) No order under section 25, 30 or 31 shall be made unless a draft of it has been laid before and approved by resolution of—

(a) each House of Parliament, if it is made by the Secretary of State or the Minister of Agriculture, Fisheries and Food;

(b) the Scottish Parliament, if it is made by the Scottish Ministers;

(c) the Northern Ireland Assembly, if it is made by the First Minister and deputy First Minister or by a Northern Ireland Department.

(4) A statutory instrument made under section 27 or 42 is subject to annulment in pursuance of a resolution of—

(a) either House of Parliament, if it is made by the Secretary of State;

(b) the Scottish Parliament, if it is made by the Scottish Ministers;

and a statutory rule made under that section is subject to negative resolution within the meaning of section 41(6) of the Interpretation Act (Northern Ireland) 1954.

(5) No recommendation shall be made to Her Majesty to make an Order in Council under section 32 or 33 unless a draft of the Order has been laid before and approved by resolution of each House of Parliament, the National Assembly for Wales, the Scottish Parliament and the Northern Ireland Assembly.

Crown application.

38.—(1) This Act binds the Crown (but does not affect Her Majesty in her private capacity).

(2) Subsection (1)—

(a) does not require subordinate legislation made under this Act to bind the Crown; and

(b) is to be interpreted as if section 38(3) of the Crown Proceedings Act 1947 (references to Her Majesty in her private capacity) were contained in this Act.

(3) If the Secretary of State certifies that it appears to him requisite or expedient in the interests of national security that the powers of entry conferred by sections 11 and 14 should not be exercisable in relation to any premises specified in the certificate, being premises held or used by or on behalf of the Crown, those powers shall not be exercisable in relation to those premises.

Financial provisions.

39.—(1) There shall be paid out of money provided by Parliament—

(a) any expenditure incurred by a Minister of the Crown by virtue of this Act;

(b) any increase attributable to this Act in the sums payable out of money so provided under any other Act.

(2) Any expenditure incurred by the Agency shall be paid out of money provided by Parliament unless it is met from money paid or appropriated under

subsection (3) (or from money which the Agency is authorised by virtue of any relevant provision to apply for the purpose).

(3) Sums may be—
  (a) paid by the National Assembly for Wales;
  (b) paid out of the Scottish Consolidated Fund; or
  (c) appropriated by Act of the Northern Ireland Assembly,
for the purpose of meeting any of the expenditure of the Agency.

(4) Any sums received by the Agency, other than—
  (a) money provided by Parliament or paid or appropriated under subsection (3);
  (b) receipts which are, by virtue of provision made by or under any enactment, payable—
    (i)   to the National Assembly for Wales;
    (ii) into the Scottish Consolidated Fund; or
    (iii) into the Consolidated Fund of Northern Ireland,
    or which would be so payable but for any relevant provision relating to those receipts; and
  (c) other receipts specified, or of a description specified, in a determination under subsection 5),
shall be paid into the Consolidated Fund.

(5) The Treasury, the National Assembly for Wales, the Scottish Ministers and the Department of Finance and Personnel for Northern Ireland acting jointly may determine that any sums received by the Agency which are specified, or of a description specified, in the determination shall (instead of being payable into the Consolidated Fund by virtue of subsection (4)) be payable to the National Assembly for Wales, into the Scottish Consolidated Fund or into the Consolidated Fund of Northern Ireland, subject to any relevant provision relating to such sums.

(6) A determination under subsection (5) may be revoked or amended by a further determination.

(7) Schedule 4 (accounts and audit) has effect.

(8) In this section—
  "enactment" means an enactment contained in an Act, an Act of the Scottish Parliament or in Northern Ireland legislation;
  "relevant provision" means—
    (a) provision made by or under any Act as to the disposal of or accounting for sums payable to the National Assembly for Wales;
    (b) provision made by or under the Scotland Act 1998 or any Act of the Scottish Parliament as to the disposal of or accounting for sums payable into the Scottish Consolidated Fund; and
    (c) provision made by or under any Act or any Northern Ireland legislation as to the disposal of or accounting for sums payable into the Consolidated Fund of Northern Ireland.

Minor and consequential amendments and repeals.

40.—(1) Schedule 5 (minor and consequential amendments) has effect.

(2) Any amendment made by Schedule 5 which extends to Scotland is to be taken as a pre-commencement enactment for the purposes of the Scotland Act 1998.

(3) The National Assembly for Wales (Transfer of Functions) Order 1999 shall have effect, in relation to any Act mentioned in Schedule 1 to the Order, as if any provision of this Act amending that Act was in force immediately before the Order came into force.

(4) The enactments mentioned in Schedule 6 are repealed to the extent specified.

(5) Her Majesty may by Order in Council direct that any amendment or repeal by this Act of any provision in the 1990 Act shall extend to any of the Channel Islands with such modifications (if any) as may be specified in the Order.

Transfer of property, rights and liabilities to the Agency.

41.—(1) The Secretary of State may make one or more schemes for the transfer to the Agency of such property, rights and liabilities of a Minister of the Crown (in this section referred to as "the transferor") as appear to him appropriate having regard to the functions conferred on the Agency by provision made by or under this Act, the 1990 Act or the 1991 Order.

(2) The power conferred by subsection (1) may also be exercised by the National Assembly for Wales, the Scottish Ministers or a Northern Ireland Department in relation to their property, rights and liabilities.

(3) A transfer scheme—

    (a) may provide for the transfer of property, rights and liabilities that would not otherwise be capable of being transferred or assigned;

    (b) may define property, rights and liabilities by specifying or describing them or by referring to all of the property, rights and liabilities comprised in a specified part of the undertaking of the transferor (or partly in one way and partly in the other);

    (c) may provide for the creation—

        (i) in favour of the transferor, or of the Agency, of interests in, or rights over, property to be transferred or, as the case may be, retained by the transferor; or

        ii) of new rights and liabilities as between the Agency and the transferor;

    (d) may require the transferor or the Agency to take any steps necessary to secure that the transfer of any foreign property, rights or liabilities is effective under the relevant foreign law; and

    (e) may make such incidental, supplemental and consequential provision as the authority making it considers appropriate.

(4) On the date appointed by a transfer scheme the property, rights and liabilities which are the subject of the scheme shall, by virtue of this subsection,

become property, rights and liabilities of the Agency (and any other provisions of the scheme shall take effect).

(5) The authority making a transfer scheme may, at any time before the date so appointed, modify the scheme.

Power to make transitional provision etc.

42.—(1) The Secretary of State may by regulations make such transitional and consequential provisions and such savings as he considers necessary or expedient in preparation for, or in connection with, or in consequence of—

    (a) the coming into force of any provision of this Act; or

    (b) the operation of any enactment repealed or amended by a provision of this Act during any period when the repeal or amendment is not wholly in force.

(2) Such regulations may make modifications of any enactment (including an enactment contained in this Act).

(3) The power to make regulations under this section is also exercisable—

    (a) by the Scottish Ministers, in relation to provision that would be within the legislative competence of the Scottish Parliament to make;

    (b) by the First Minister and deputy First Minister acting jointly, in relation to provision dealing with transferred matters (within the meaning of section 4(1) of the Northern Ireland Act 1998).

Short title, commencement and extent.

43.—(1) This Act may be cited as the Food Standards Act 1999.

(2) This Act (apart from this section and paragraph 6(2) and (5) of Schedule 5) shall come into force on such day as the Secretary of State may by order appoint; and different days may be appointed for different purposes.

(3) The provisions of this Act shall be treated for the purposes of section 58 of the 1990 Act (territorial waters and the continental shelf) as if they were contained in that Act.

(4) Until the day appointed under section 3(1) of the Northern Ireland Act 1998, this Act has effect with the substitution—

    (a) for references to the First Minister and deputy First Minister acting jointly, of references to a Northern Ireland Department;

    (b) for references to an Act of the Northern Ireland Assembly, of references to a Measure of the Northern Ireland Assembly; and

    (c) for references to transferred matters within the meaning of section 4(1) of the Northern Ireland Act 1998, of references to transferred matters within the meaning of section 43(2) of the Northern Ireland Constitution Act 1973;

    (d) for references to paragraph 1(a) of Schedule 2 to the Northern Ireland Act 1998, of references to paragraph 1(a) of Schedule 2 to the Northern Ireland Constitution Act 1973.

(5) This Act extends to Scotland and Northern Ireland.

Schedules

Schedule 1 Constitution etc of the Agency

[*This Schedule is not reproduced in this work.*]

Schedule 2 Advisory Committees

[*This Schedule is not reproduced in this work.*]

Schedule 3 The Agency's Functions under other Enactments

Part I Functions under the 1990 Act

1. This Part has effect for conferring functions under the 1990 Act on the Agency (and references to sections are to sections of the 1990 Act).

Section 6 (enforcement)

2. The Agency—
   (a) may be directed to discharge duties of food authorities under section 6(3);
   (b) may be specified as an enforcement authority for regulations or orders in pursuance of section 6(4); and
   (c) may take over the conduct of proceedings mentioned in section 6(5) either with the consent of the person who instituted them or when directed to do so by the Secretary of State.

Section 13(3) (emergency control orders)

3. The Agency may grant consent under subsection (3), and give directions under subsection (5), of section 13.

Section 40 (codes of practice)

4.—(1) The Agency may, after consulting the Secretary of State—
   (a) give directions to food authorities under section 40(2)(b) as to steps to be taken in order to comply with codes of practice under section 40; and

(b) enforce any such directions.

(2) The Agency may undertake consultation with representative organisations regarding proposals for codes of practice under section 40.

Section 41 (information from food authorities)

5. The Agency may exercise the power to require returns or other information from food authorities under section 41.

Section 42 (default powers)

6. The Agency may be empowered by an order under section 42 to discharge any duty of a food authority.

Section 48 (regulations and orders)

7. The Agency may undertake consultation with representative organisations required by section 48 regarding proposals for regulations or orders under the 1990 Act.

Part II Functions under the 1991 Order

8. This Part has effect for conferring functions under the 1991 Order on the Agency (and references to Articles are to Articles of the 1991 Order).

Article 12 (emergency control orders)

9. The Agency may grant consent under paragraph (3), and give directions under paragraph (5), of Article 12.

Article 26 (enforcement)

10. The Agency—
    (a) may be directed to discharge duties of district councils under Article 26(2);
    (b) may be specified as an authority to enforce and execute regulations or orders in pursuance of Article 26(3); and
    (c) may take over the conduct of proceedings mentioned in Article 26(4) either when directed to do so by the Department of Health and Social Services for Northern Ireland or with the consent of the district council which instituted them.

Article 39 (codes of practice)

11.—(1) The Agency may, after consulting the Department of Health and Social Services for Northern Ireland—
  (a) give directions to district councils under Article 39(2)(b) as to steps to be taken in order to comply with codes of practice under Article 39; and
  (b) enforce any such directions.
(2) The Agency may undertake consultation with representative organisations regarding proposals for codes of practice under Article 39.

Article 40 (information from district councils)

12. The Agency may exercise the power to require returns or other information from district councils under Article 40.

Article 41 (default powers)

13. The Agency may be empowered by an order under Article 41 to discharge any duty of a district council.

Article 47 (regulations and orders)

14. The Agency may undertake consultation with representative organisations required by Article 47 regarding proposals for regulations or orders under the 1991 Order.

Part III Other functions

Medicines Act 1968 (c. 67)

15.—(1) The Medicines Act 1968 shall be amended as follows.
(2) In section 4 (establishment of committees), after subsection (5) there shall be inserted the following subsection—
  "(5A) Where a committee is established under this section for purposes including the consideration of veterinary products as defined in section 29(2) of the Food Standards Act 1999, one member of the committee shall be appointed by the Ministers establishing the committee on the nomination of the Food Standards Agency."
(3) In section 129 (orders and regulations), after subsection (6) there shall be inserted the following subsection—
  "(6A) The organisations to be consulted under subsection (6) of this section include, where any provisions of the regulations or order apply to

veterinary products as defined in section 29(2) of the Food Standards Act 1999, the Food Standards Agency."

Food and Environment Protection Act 1985 (c. 48)

16.—(1) The Agency shall have the following functions under the Food and Environment Protection Act 1985.

(2) The Agency may exercise the following powers under section 2 (powers when emergency order has been made)—

(a) the power to give consents under subsection (1);

(b) the power to give directions or do anything else under subsection (3);

(c) the power to recover expenses under subsection (5) or (6).

(3) In section 7 (exemptions from need for licence under Part II), after subsection (3) there shall be inserted the following subsection-

"(3A) A licensing authority—

(a) shall consult the Food Standards Agency as to any order the authority contemplates making under this section; and

(b) shall from time to time consult that Agency as to the general approach to be taken by the authority in relation to the granting of approvals and the imposition of conditions under subsections (2) and (3) (including the identification of circumstances in which it may be desirable for the Agency to be consulted in relation to particular cases)."

(4) In section 8 (licences under Part II), after subsection (11) there shall be inserted the following subsections-

"(11A) The matters to which a licensing authority is to have regard in exercising powers under this section include any advice or information given to that authority by the Food Standards Agency (whether of a general nature or in relation to the exercise of a power in a particular case).

(11B) A licensing authority shall from time to time consult the Food Standards Agency as to the general manner in which the authority proposes to exercise its powers under this section in cases involving any matter which may affect food safety or other interests of consumers in relation to food (including the identification of circumstances in which it may be desirable for the Agency to be consulted in relation to particular cases)."

(5) In section 16 (control of pesticides), after subsection (9) there shall be inserted the following subsection-

"(9A) The Ministers—

(a) shall consult the Food Standards Agency as to regulations which they contemplate making; and

(b) shall from time to time consult that Agency as to the general approach to be taken by them in relation to the giving, revocation or suspension of approvals and the imposition of conditions on approvals (including the identification of circumstances in which it may be desirable for the Agency to be consulted in relation to particular cases)."

(6) In Schedule 5 (the Advisory Committee), after paragraph 1 there shall be inserted the following paragraph—

"1A. The committee shall include one member appointed by the Ministers on the nomination of the Food Standards Agency."

Environmental Protection Act 1990 (c. 43)

17. In section 108(7) and section 111(7) of the Environmental Protection Act 1990 (grant of exemptions) after the words "Secretary of State" there shall be inserted the words ", or by the Secretary of State and the Food Standards Agency acting jointly,".

18. For section 126 of that Act (exercise of certain functions relating to genetically modified organisms jointly by Secretary of State and Minister of Agriculture, Fisheries and Food) there shall be substituted the following section—

"Mode of exercise of certain functions.

126.—(1) Any power of the Secretary of State to make regulations under this Part (other than the power conferred by section 113 above) is exercisable, where the regulations to be made relate to any matter with which the Minister is concerned, by the Secretary of State and the Minister acting jointly.

(2) Any function of the Secretary of State under this Part (other than a power to make regulations) is exercisable, where the function is to be exercised in relation to a matter with which the Minister is concerned, by the Secretary of State and the Minister acting jointly (but subject to subsection (3) below).

(3) Any function of the Secretary of State under sections 108(8) and 110 above is exercisable, where the function is to be exercised in relation to a matter with which the Agency is concerned—

(a) if it is a matter with which the Minister is also concerned, by the Secretary of State, the Minister and the Agency acting jointly;

(b) otherwise, by the Secretary of State and the Agency acting jointly.

(4) Accordingly, references in this Part to the Secretary of State shall, where subsection (1), (2) or (3) above applies, be treated as references to the authorities in question acting jointly.

(5) The Agency shall be consulted before—

(a) any regulations are made under this Part, other than under section 113 above, or

(b) any consent is granted or varied.

(6) The reference in section 113 above to expenditure of the Secretary of State in discharging functions under this Part in relation to consents shall be taken to include a reference to the corresponding expenditure of the Minister in discharging those functions jointly with the Secretary of State.

(7) The validity of anything purporting to be done in pursuance of the exercise of a function of the Secretary of State under this Part shall not be affected by any question whether that thing fell, by virtue of this section, to be done jointly with the Minister or the Agency (or both).

(8) In this section—

"the Agency" means the Food Standards Agency; and

"the Minister" means the Minister of Agriculture, Fisheries and Food."

Genetically Modified Organisms (Northern Ireland) Order 1991 (S.I. 1991/1714 (N.I. 19))

19. In Article 5(7) and Article 8(7) of the Genetically Modified Organisms (Northern Ireland) Order 1991 (grant of exemptions) after the word "Department" there shall be inserted the words ", or by the Department and the Food Standards Agency acting jointly,".

20.—(1) For Article 22 of that Order (exercise of certain functions relating to genetically modified organisms jointly by the Department of the Environment and the Department of Agriculture) there shall be substituted the following Article—

"Mode of exercise of certain functions

22.—(1) Any power of the Department to make regulations under this Order (other than the power conferred by Article 10) is exercisable, where the regulations to be made relate to any matter with which the Department of Agriculture is concerned, by the Department and the Department of Agriculture acting jointly.

(2) Any function of the Department under this Order (other than a power to make regulations) is exercisable, where the function is to be exercised in relation to a matter with which the Department of Agriculture is concerned, by the Department and the Department of Agriculture acting jointly (but subject to paragraph (3)).

(3) Any function of the Department under Articles 5(8) and 7 is exercisable, where the function is to be exercised in relation to a matter with which the Food Standards Agency is concerned—

    (a) if it is a matter with which the Department of Agriculture is also concerned, by the Department, the Department of Agriculture and the Food Standards Agency acting jointly;

(b) otherwise, by the Department and the Food Standards Agency acting jointly.

(4) Accordingly, references in this Order to the Department shall, where paragraph (1), (2) or (3) applies, be treated as references to the authorities in question acting jointly.

(5) The Food Standards Agency shall be consulted before—

(a) any regulations are made under this Order, other than under Article 10, or

(b) any consent is granted or varied.

(6) The reference in Article 10 to expenditure of the Department in discharging functions under this Order in relation to consents shall be taken to include a reference to the corresponding expenditure of the Department of Agriculture in discharging those functions jointly with the Department.

(7) The validity of anything purporting to be done in pursuance of the exercise of a function of the Department under this Order shall not be affected by any question whether that thing fell, by virtue of this Article, to be done jointly with the Department of Agriculture or the Food Standards Agency (or both)."

(2) In consequence of sub-paragraph (1), in the definition of "the Department" in Article 2(2) of that Order, after the word "means" there shall be inserted the words "(subject to Article 22)".

Radioactive Substances Act 1993 (c. 12)

21. The Agency shall have the right to be consulted in the circumstances mentioned in subsection (4A) of section 16 or subsection (2A) of section 17 of the Radioactive Substances Act 1993 (proposals for granting or varying authorisations) about the matters mentioned in paragraphs (a) and (b) of that subsection.

Schedule 4 Accounts and Audit

[*This Schedule is not reproduced in this work.*]

Schedule 5 Minor and Consequential Amendments

[*This Schedule is not reproduced in this work.*]

Schedule 6 Repeals

[This Schedule is not reproduced in this work.]

· Index ·